The Student Edition of SIMULINK®

Dynamic System Simulation for MATLAB®

The Student Edition of SIMULINK®

Dynamic System Simulation for MATLAB®

James B. Dabney
Thomas L. Harman

User's Guide

The MATLAB® Curriculum Series

Library of Congress Cataloging-in-Publication Data
The Student Edition of SIMULINK : version 2, user's guide / The MathWorks, Inc.: by James Dabney
p. cm. — (The MATLAB curriculum series)
Includes bibliographical references (p. –) and index.
ISBN 0-13-659699-1 (user's guide)
1. SIMULINK 2. Numerical analysis—data processing
I. MathWorks, Inc. II. Series.
QA297.58436 1998
519.4'1285'53042—dc21 98-11102
CIP

Publisher: Tom Robbins
Editorial / Production Supervision: Rose Kernan
Editor-in-Chief: Marcia Horton
Managing Editor: Bayani Mendoza de Leon
Vice President of Production and Manufacturing: David W. Riccardi
Manufacturing Buyer: Donna Sullivan
Manufacturing Manager: Trudy Pisciotti
Editorial Assistant: Nancy Garcia
Book Cover and Box Design: The MathWorks, Inc.
CD-Holder Design: Paul Gourhan
Composition: RDD Consultants, Inc.

Printed in the United States of America

10 9 8 7 6 5 4 3 2

ISBN 0-13-659699-1

Prentice-Hall International (UK) Limited, *London*
Prentice-Hall of Australia Pty. Limited, *Sydney*
Prentice-Hall Canada, Inc., *Toronto*
Prentice-Hall Hispanoamericana, S.A., *Mexico*
Prentice-Hall of India Private Limited, *New Delhi*
Prentice-Hall of Japan, Inc., *Tokyo*
Simon & Schuster Asia Pte. Ltd., *Singapore*
Editora Prentice-Hall do Brasil, Ltda., *Rio de Janeiro*

The MathWorks, Inc.
24 Prime Park Way
Natick, Massachusetts 01760-1500
Phone: (508) 647-7000
Fax: (508) 647-7001
E-mail: info@mathworks.com
http://www.mathworks.com

Contents

Preface

To the Instructor

SIMULINK is a package for use with MATLAB for modeling, simulating, and analyzing dynamical systems. Its graphical modeling environment uses familiar block diagrams, so systems illustrated in texts can be easily implemented in SIMULINK. The simulation is interactive, so you can change parameters and immediately see what happens. The analysis tools include those built into SIMULINK, plus the many tools in MATLAB and its application toolboxes. This combination of ease of use with flexible and powerful capability has already made SIMULINK the choice of thousands of engineers, instructors, and students in industry and academia.

Now, The Student Edition of SIMULINK, in combination with The Student Edition of MATLAB, gives students an affordable way to use this powerful modeling and simulation environment in their studies, while learning a tool that will prove invaluable throughout their careers.

SIMULINK allows students to move beyond idealized linear models, to explore more realistic nonlinear models that account for friction, air resistance, gear slippage, and other real-world phenomena. It turns the student's computer into a virtual laboratory for doing detailed analysis of systems that simply wouldn't be possible or practical otherwise. These systems might describe the response of an electric motor, the flight dynamics of an airplane, the active suspension of a car, or the effect of the monetary supply on the economy. And SIMULINK makes such analysis fun.

The Student Edition of SIMULINK for the student's own personal computer is an excellent complement to educationally discounted licenses of the professional version, such as computer laboratory licenses or workstation site licenses. SIMULINK 2 models are fully compatible, both between the student and professional versions and across platforms. As a result, students can take their models to the lab to use advanced tools like code generation or real-time hardware support provided by the SIMULINK Real-Time Workshop.

By itself, or when coupled with texts, SIMULINK can be effectively incorporated into the curriculum to enhance students' modeling and analysis skills, as well as understanding of systems dynamics and behavior.

Technical Support for Instructors

The MathWorks provides technical support to registered instructors who use The Student Edition of SIMULINK in their courses. For technical questions, instructors can direct inquiries

- via WWW: `http://www.mathworks.com/support.html`
- Via e-mail: `support@mathworks.com`
- Via telephone: (508) 647-7000
- Via fax: (508) 647-7201

Other Information Sources for Instructors and Students

- Use the SIMULINK on-line help facility by typing `help simulink` at the MATLAB prompt.
- Use the Block Browser for help on any block.
- Use the MATLAB Help Desk.
- Students and instructors with access to Usenet newsgroups can participate in the `comp.soft-sys.matlab` newsgroup. Here, an active community of MATLAB and SIMULINK users—spanning industries, countries, applications, and schools—exchange ideas, help with each other's questions and problems, and share user-written functions and tools. Members of The MathWorks staff also participate, and the newsgroup has become a stimulating, open, and free-flowing forum.
- On the World Wide Web (WWW), use Netscape Navigator or another browser to reach The MathWorks Home Page at `http://www.mathworks.com`.
- The MathWorks maintains an electronic archive of user-contributed routines, product information, and other useful things. It can be reached using anonymous ftp to `ftp.mathworks.com`, or from The MathWorks Home Page on the WWW.
- The quarterly MathWorks newsletter *MATLAB News & Notes* provides information on new products, technical notes and tips, application articles, a calendar of trade shows and conferences, and other useful information. *MATLAB News & Notes* is free to registered users of The Student Edition of SIMULINK.

MATLAB based Books

A number of books can be used with the student editions of MATLAB and SIMULINK, many featuring exercises, problems sets, and supplemental functions. These include standard texts or supplemental workbooks in a broad range of courses, such as Control Theory, Signals and Systems, and Linear Systems.

For a current list of MATLAB based books, consult The MathWorks Home Page on the WWW at `http://www.mathworks.com`.

Acknowledgments

The Student Edition of SIMULINK 2 is the product of a collaborative effort between The MathWorks and Prentice Hall, with many people contributing to its development. At The MathWorks, we especially want to acknowledge Rick Spada, Howie Taitel, Bob Gilmore, Carlie Schubert, Liz Callanan, Scott Gray, Naomi Bulock and Jim Tung. At Prentice Hall, there have been contributions from Marcia Horton, Tom Robbins, Nancy A. Garcia, Joe Scordato, Rose Kernan, Gary June, Amy Rosen, Paula Maylahn, Joe Sengotta, and Donna Sullivan.

1

Getting Started

This chapter introduces SIMULINK and explains how you can install The Student Edition of SIMULINK 2 on your Windows or Macintosh computer.

1.1 To the Student

Welcome to The Student Edition of SIMULINK 2! In the last few years, SIMULINK has become the most widely used software package for modeling and simulating dynamical systems in academia and industry. Now, The Student Edition of SIMULINK 2, in combination with The Student Edition of MATLAB 5, makes it practical for you to use this powerful environment on your own personal computer in your home, dorm, or wherever you study.

SIMULINK encourages you to try things out. You can easily build models from scratch or take an existing model and add to it. Simulations are interactive, so you can change parameters "on the fly" and immediately see what happens. You have instant access to all of the analysis tools in MATLAB, so you can take the results and analyze and visualize them. We hope that you will get a sense of the *fun* of modeling and simulation, through an environment that encourages you to pose a question, model it, and see what happens.

With SIMULINK, you can move beyond idealized linear models to explore more realistic nonlinear models, factoring in friction, air resistance, gear slippage, hard stops, and the other things that describe real-world phenomena. SIMULINK turns your computer into a lab for modeling and analyzing systems that simply wouldn't be possible or practical otherwise, whether the behavior of an automobile clutch system, the flutter of an airplane wing, the dynamics of a predator-prey model, or the effect of the monetary supply on the economy.

It is also practical. With thousands of engineers around the world using SIMULINK to model and solve real problems, knowledge of these tools will serve you well, not only in your studies but also throughout your professional career.

We hope you enjoy exploring the software.

1.1.1 What Is SIMULINK?

SIMULINK is a software package for modeling, simulating, and analyzing dynamical systems. It supports linear and nonlinear systems, modeled in continuous time, sampled time, or a hybrid of the two. Systems can also be multirate (i.e., have different parts that are sampled or updated at different rates).

For modeling, SIMULINK provides a graphical user interface (GUI) for building models as block diagrams, using click-and-drag mouse operations. With this interface, you can draw the models just as you would with pencil and paper (or as most textbooks depict them). This is a far cry from previous simulation packages that require you to formulate differential equations or difference equations in a language or program. SIMULINK includes a comprehensive block library of sinks, sources, linear and nonlinear components, and connectors. You can also customize and create your own blocks.

Models are hierarchical, so you can build models using both top-down and bottom-up approaches. You can view the system at a high level, then double click on blocks to go down through the levels to see increasing levels of model detail. This provides insight into how a model is organized and how its parts interact.

After you define a model, you can simulate it, using a choice of solvers, either from the SIMULINK menus or by entering commands in the MATLAB command window. The menus are particularly convenient for interactive work, while the command line approach is very useful for running a batch of simulations (for example, if you are doing Monte Carlo simulation or want to sweep a parameter across a range of values). Using scopes and other display blocks, you can see the simulation results while the simulation is running. In addition, you can change parameters and immediately see what happens, for "what if" exploration. The simulation results can be put in the MATLAB workspace for postprocessing and visualization.

Model analysis tools include linearization and trimming tools that can be accessed from the MATLAB command line, plus the many tools in MATLAB and its application toolboxes. And because MATLAB and SIMULINK are integrated, you can simulate, analyze, and revise your models in either environment at any point.

1.1.2 What's New in SIMULINK 2?

SIMULINK 2 provides many improvements designed to make model building easier. Other changes can make models execute faster and more reliably. Among the many improvements are the following:

- New, more intuitive user interface
- On-line Block Browser providing easy access to detailed documentation for all SIMULINK blocks
- Improved automatic signal line routing
- Signal line labels
- Signal line labels that propagate through connecting blocks
- Automatic labeling of subsystem ports
- Enhancements to many blocks
- New Scope block providing vastly improved display of signals
- New dialog box for setting simulation parameters
- Improved commands to run SIMULINK models from the MATLAB command line
- Greatly improved masking capabilities
- Conditionally executed subsystems
- New differential equation solvers providing faster, more accurate simulation results
- Zero crossing detection
- Improved handling of algebraic loops

1.1.3 How to Use this Manual

This manual contains eight chapters and an appendix. Chapters 2 through 7 describe the procedures and techniques you'll need to build SIMULINK models. Chapter 8 provides useful information on numerical issues. The Appendix lists all the standard SIMULINK blocks.

Chapter 2 is intended to get you started using SIMULINK. The chapter shows you how to start SIMULINK, build a simple model, run the model, and save it. It also introduces the Block Browser, the new on-line block documentation system for SIMULINK. Unless you're already familiar with SIMULINK, you'll want to work through the tutorial examples in Chapter 2, since the skills you'll acquire will be needed in the following chapters. Because SIMULINK is graphical and interactive, we encourage you to jump right in and try it.

Chapter 3 covers all the mechanics of building and running a SIMULINK model. It also discusses how to print a model.

Chapter 4 explains how to use SIMULINK to model continuous systems. The chapter covers simple scalar continuous systems, transfer functions, and vector models and discusses modeling nonlinear continuous systems.

Chapter 5 discusses discrete systems, in a manner that builds on the continuous systems discussion in Chapter 4. The chapter also covers hybrid systems, which contain both continuous and discrete components.

Chapter 6 is devoted to subsystems and masking. Subsystems are the SIMULINK equivalent to subprograms. Masking allows you to convert SIMULINK subsystems into custom blocks, complete with icons that indicate the blocks' function and on-line help screens.

Chapter 7 provides the information you'll need to use the SIMULINK analysis tools. The chapter shows you how to run SIMULINK models from the MATLAB command line or from within MATLAB programs. The linearization tools allow you to linearize a SIMULINK model about any point in its state space, and the trim tools allow you to locate equilibria.

Chapter 8 discusses numerical issues. SIMULINK provides a wide selection of differential equation solvers, including some very advanced algorithms for solving stiff systems. The chapter discusses the issues you need to consider in selecting the best solver for your models. It also discusses algebraic loops, an important numerical issue that can affect both the speed and accuracy of a simulation.

The Appendix provides a tabular listing of all the blocks in the SIMULINK block libraries. Detailed block documentation is available on-line via the Block Browser.

Although we have tried to provide the most complete and up-to-date information in this manual, some information may have changed after it was printed. Please check the `README` file delivered with your SIMULINK system for the latest release notes.

1.1.4 Comparing The Student Edition to Professional SIMULINK

The Student Edition of SIMULINK 2 is available for Windows 95 and Windows NT compatible personal computers and Macintosh systems. It is identical to the SIMULINK 2.0 professional version except for the following:

- It requires The Student Edition of MATLAB 5.
- Models are limited to 50 blocks. (Note that the Subsystem block and Inport, Outport, Mux, Demux, Selector, Goto, and From blocks are not included in this count, so there is no penalty for making your model hierarchical. However, the blocks within a subsystem are included in this count. Some

SIMULINK blocks are masked subsystems and contain more than one block.)

- It is available in single-user licenses only (no networking).

The Student Edition of SIMULINK provides a Student User Upgrade Discount for purchase of the professional version (refer to the registration card for more information).

1.1.5 Upgrading to Professional SIMULINK

The professional versions of MATLAB and SIMULINK are available for Windows 95 and NT and Macintosh personal computers; UNIX workstations from Sun, Hewlett-Packard, IBM, Silicon Graphics, and DEC; and Open VMS computers. For product information or to place an order, call or write your educational account representative at The MathWorks:

The MathWorks, Inc.
University Sales Department
24 Prime Park Way
Natick, MA 01760-1500
Phone: (508) 647-7000
Fax: (508) 647-7001
E-mail: `info@mathworks.com`
WWW: `http://www.mathworks.com`

1.1.6 Technical Support

If you have problems with SIMULINK, there are a variety of sources of help. This section discusses getting help, registering SIMULINK, replacing defective disks, and using the limited warranty.

Student Support Policy

Neither Prentice Hall, Inc., nor The MathWorks, Inc., provides technical support directly to student users of The Student Edition of SIMULINK. If you encounter difficulty while using the Student Edition software,

1 Read the relevant tutorial and reference section of this *User's Guide* containing information on the commands or procedures you are trying to execute.

2 Use the software's on-line help facility by typing `help simulink` at the MATLAB prompt. Additionally, the Block Browser (described in Chapter 2) provides details on all the blocks in the SIMULINK block library.

3 Write down the sequence of procedures you were executing so that you can explain the nature of the problem to your instructor. Be certain to note the exact error message you encountered.

4 If you have Internet access, refer to The MathWorks Web site at `http://www.mathworks.com`. The Web site contains lots of useful information and is updated regularly. The site also provides a Solution Search capability that queries The MathWorks extensive database of technical support information.

5 If you have consulted this *User's Guide* and the on-line help and are still stymied, you can post your question to the `comp.soft-sys.matlab` usegroup, if you have access to Usenet newsgroups. Many active SIMULINK users participate in the newsgroup, and they are a good resource for answers or tips about using SIMULINK.

Student User Registration

Students who have purchased the software package will find a card in the package for registering as a user of The Student Edition of SIMULINK. Take a moment now to complete and return this card. Registered student users

- Are entitled to replace defective disks at no charge.
- Qualify for a discount on upgrades to professional versions of SIMULINK.
- Receive the *MATLAB News & Notes* quarterly newsletter, which contains information on new products, technical notes and tips, application articles, a calendar of trade shows and conferences, and other useful information.
- Become active members of the worldwide SIMULINK user community.

Defective CD Replacement

E-mail or fax your request for CD replacement to Prentice Hall at `matlab@prenhall.com` or (201)236-7170. You must send us your damaged or defective CD, and we will provide you with a new one.

Limited Warranty

No warranties, express or implied, are made by The MathWorks, Inc. that the program or documentation is free of error. Further, The MathWorks, Inc. does not warrant the program for correctness, accuracy, or fitness for a task. You rely on the results of the program solely at your own risk. The program should not be relied on as the sole basis to solve a problem whose incorrect solution could result in injury to person or property. If the program is employed in such a manner, it is at the user's own risk, and The MathWorks, Inc. disclaims all liability for such misuse. Neither the MathWorks, Inc. nor anyone else who has been involved in the creation, production, or delivery of this program shall be liable for any direct or indirect damages.

1.2 SIMULINK for Windows 95/NT

The Student Edition of SIMULINK 2 for Windows 95 and Windows NT takes advantage of the Windows 32-bit architecture. It is not compatible with earlier versions of Windows, such as Windows 3.1.

1.2.1 System Requirements

The Student Edition of SIMULINK 2 for Windows has the same system requirements as The Student Edition of MATLAB 5. The Windows version of The Student Edition of SIMULINK 2 requires

- The Student Edition of MATLAB 5 for Microsoft Windows
- A personal computer compatible with Windows 95 or Windows NT (80386 DX, 80486, or Pentium processor)
- Windows 95 or Windows NT (3.51 or 4.0)
- 8 MB of memory (16 MB recommended)
- 20 MB of free disk space (after installing MATLAB)
- A CD-ROM drive
- A Microsoft Windows-supported mouse and monitor

These items are recommended:

- Additional memory
- A Microsoft Windows-supported graphics accelerator card
- A Microsoft Windows-supported printer

1.2.2 Installing SIMULINK for Windows 95/NT

The Student Edition of SIMULINK 2 for Windows ships on a single CD-ROM. Before installing SIMULINK, be sure that The Student Edition of MATLAB 5 is installed and running properly. When you are ready to install The Student Edition of SIMULINK, follow these steps:

1 Start Windows.

2 Insert the SIMULINK CD into your CD-ROM drive. The installation program may run automatically at this time.

3 If the installation program does not run automatically, open the Windows Explorer, and then change to the CD-ROM directory and double click on `setup.exe`.

4 Accept or reject the software licensing agreement. If you accept, the installation will proceed.

The installation program takes a few minutes to run as the files are copied to your hard disk. After installation, there will be a `simulink` directory in the root MATLAB directory, and a `simulink` directory in the MATLAB `toolbox` directory.

1.3 SIMULINK for Macintosh

The Student Edition of SIMULINK 2 for Macintosh is compatible with most Macintosh computers, including Power Macintosh.

1.3.1 System Requirements

The Student Edition of SIMULINK 2 for Macintosh has the same system requirements as The Student Edition of MATLAB 5. The Macintosh version requires

- The Student Edition of MATLAB 5 for Macintosh
- A Macintosh computer (Power PC, 68020 or 68030 with 68881 or 68882 math coprocessor, or 68040)
- 16 MB of memory
- 20 MB of free disk space (after installing MATLAB)
- A CD-ROM drive
- A mouse and monitor

1.3.2 Installing SIMULINK for Macintosh

The Student Edition of SIMULINK 2 for Macintosh ships on a single CD-ROM. Before installing SIMULINK, be sure that The Student Edition of MATLAB 5 is installed and running properly. When you are ready to install The Student Edition of SIMULINK, follow these steps:

1 Insert the SIMULINK CD into your CD-ROM drive. Double click on the `Install simulink` icon.

2 Accept or reject the software licensing agreement. If you accept, the installation will proceed.

The installation program takes a few minutes to run as the files are copied to your hard disk. After installation, there should be a `simulink` directory in the root MATLAB directory, and a `simulink` directory in the MATLAB `toolbox` directory.

2

Quick Start

In this chapter, we will discuss the basics of building and executing SIMULINK models. We will start with a simple first-order system. Next, we will build a more complex model that includes feedback and illustrates several important procedures in SIMULINK programming. Finally, we will discuss the SIMULINK help system.

2.1 Introduction

SIMULINK is a very powerful programming language, and a big part of that power is its ease of use. In this chapter, we will introduce SIMULINK programming by building and executing two simple models. Our purpose here is to cover the basics of model building and execution. We will discuss these topics in more detail in later chapters.

2.1.1 Typographical Conventions

Before we build the first model, we need to establish some typographical conventions.

Computer Type

All computer input, output, variable names, and command names are shown in `sans-serif` font.

Using Menus

The SIMULINK user interface is based on four pull-down menus located on a menu bar at the top of the SIMULINK window. To facilitate discussion of the various menu choices, we will use the convention [**Menu bar choice**]: [**Pull-down menu choice**]. For example, choosing "File" followed by "Save As" (as shown in Figure 2-1) will be written **File:Save As**.

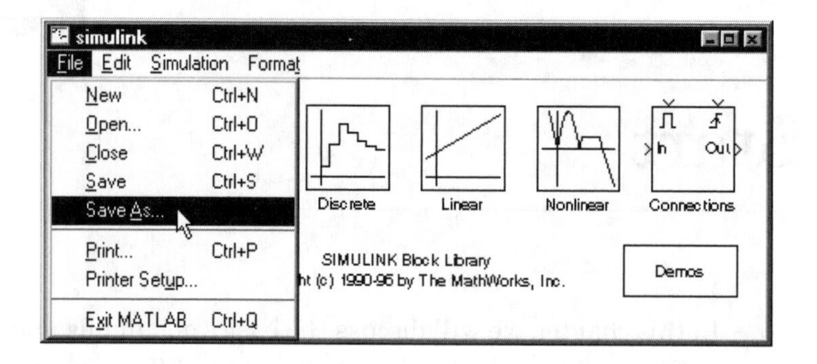

Figure 2-1 Menu selection

Dialog Box Fields

SIMULINK makes extensive use of dialog boxes to set simulation parameters and configure blocks. Dialog box fields are indicated in **bold** type. For example, choosing **File:Save As** opens the dialog box shown in Figure 2-2. The fields in this dialog box are **Save in**, **File name**, and **Save as type**.

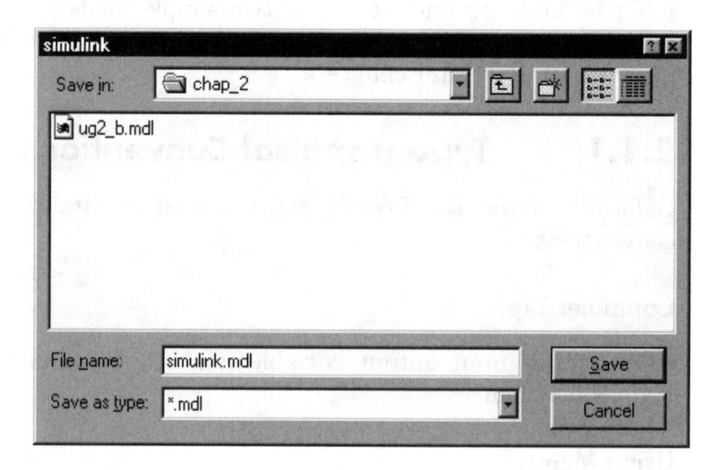

Figure 2-2 **File:Save As** dialog box

2.2 Building a Simple Model

Let's start by building a SIMULINK model that solves the differential equation

$$\dot{x} = \sin(t), \tag{2-1}$$

where $x(0) = 0$.

SIMULINK is an extension to MATLAB, and it must be invoked from within MATLAB. Start SIMULINK by clicking on the SIMULINK icon on the MATLAB tool bar, as shown in Figure 2-3. An alternative is to enter the command `simulink` at the MATLAB prompt.

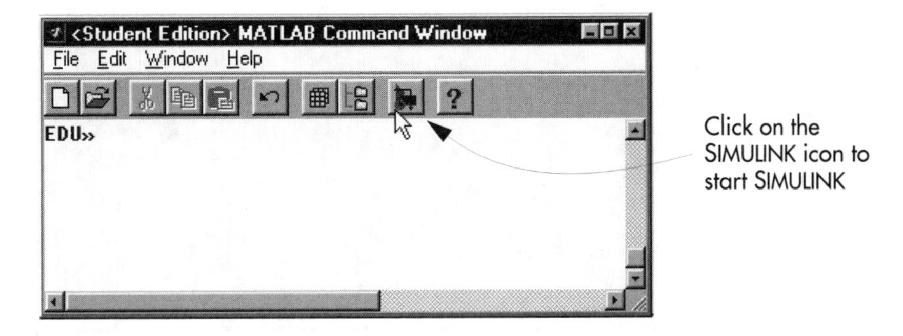

Figure 2-3 Starting SIMULINK

Two new windows will appear on the screen. The first window is the SIMULINK block library, shown in Figure 2-4. The second window (Figure 2-5) is an empty model window, named `untitled`, in which you will build the SIMULINK model.

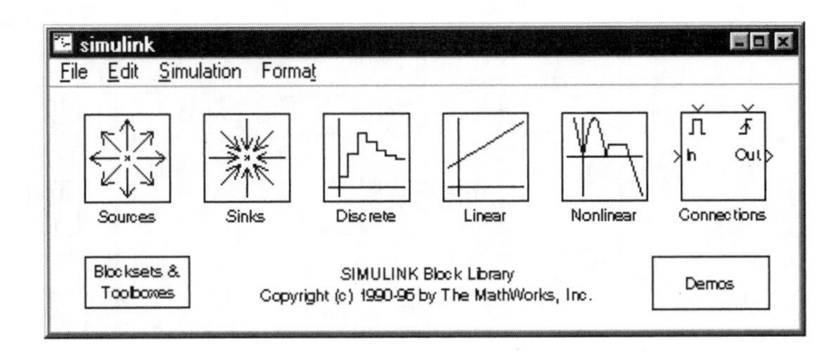

Figure 2-4 SIMULINK block library window

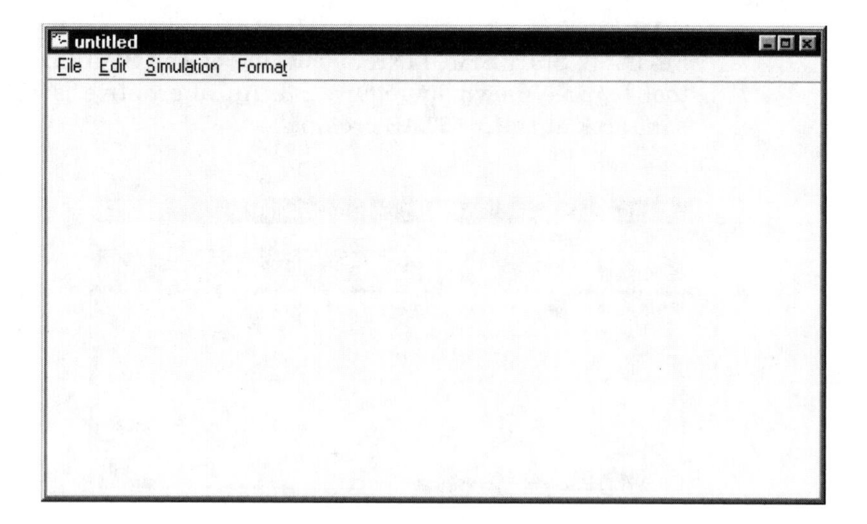

Figure 2-5 Empty model window

Double click on the Sources icon in the SIMULINK block library, opening the Sources block library.

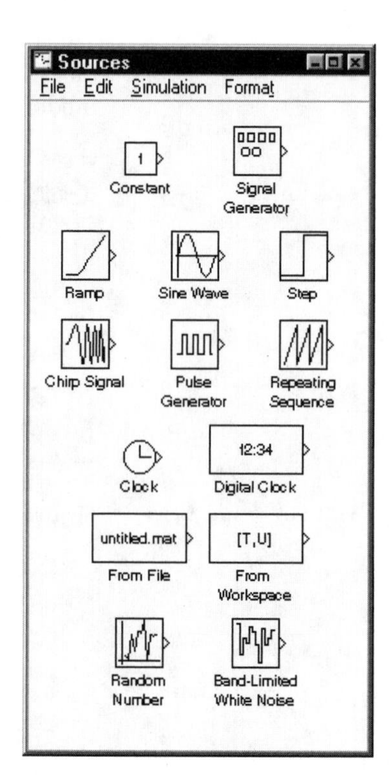

Drag the Sine Wave block from the Sources block library to the model window, positioning it as shown. A copy of the block is placed in the model window. Note that in the figure, we have resized the model window to save space.

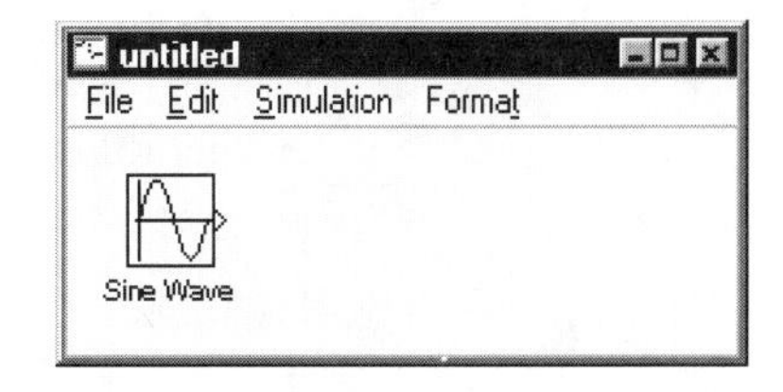

Open the Linear block library and drag an Integrator block to the model window.

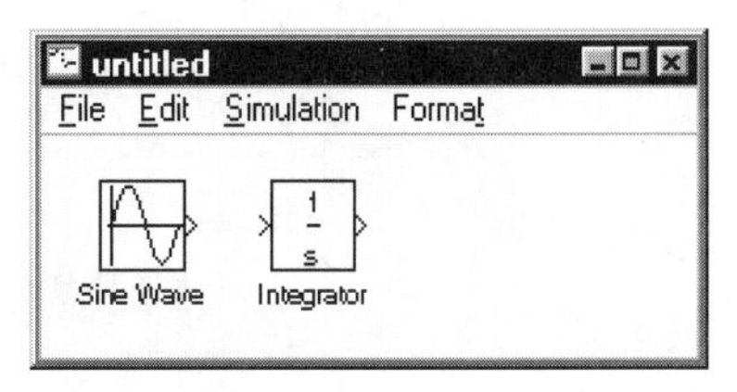

Open the Sinks block library and drag a Scope block to the model window.

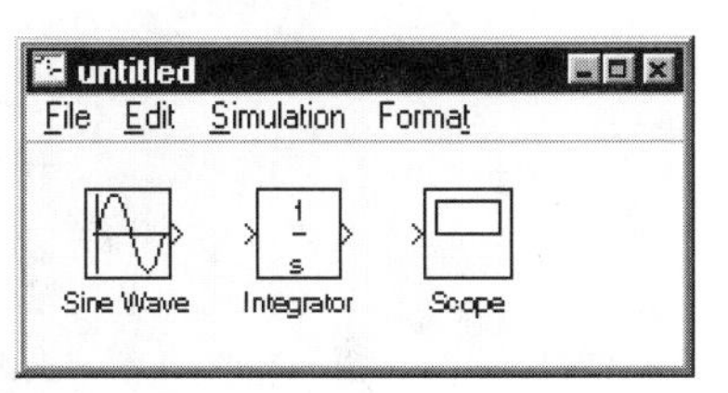

Next, connect the blocks with signal lines to complete the model:

Place the cursor on the output port of the Sine Wave block. The output port is the > symbol on the right edge of the block. The cursor changes to a cross-hair shape when it's on the output port.

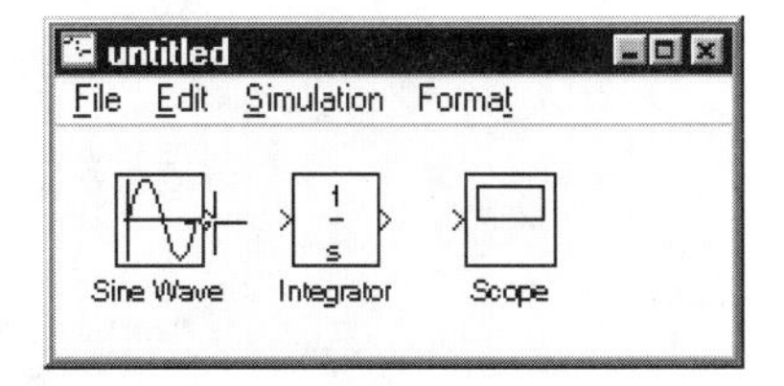

Drag from the output port to the input port of the Integrator block. The input port is the > symbol on the left edge of the block. As you drag, the cursor retains the cross-hair shape.

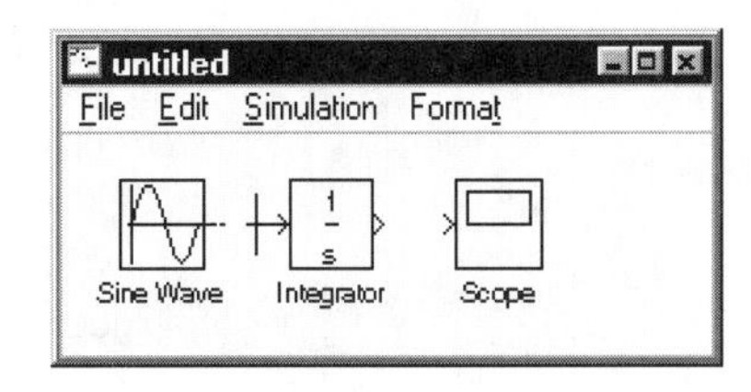

Now the model should look like this. The signal line has an arrowhead indicating the direction of signal flow.

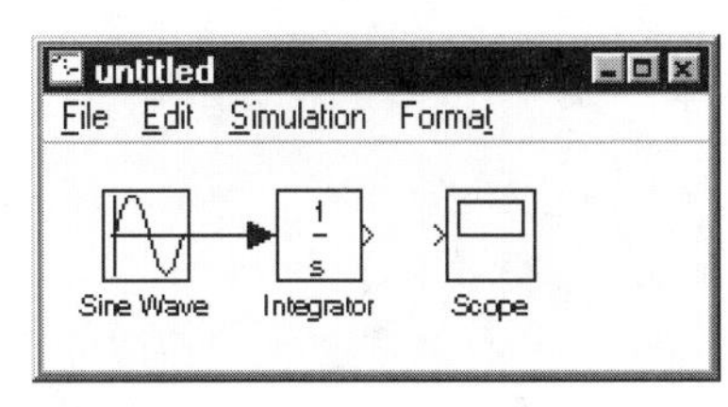

Draw another signal line from the Integrator output port to the Scope input port, completing the model.

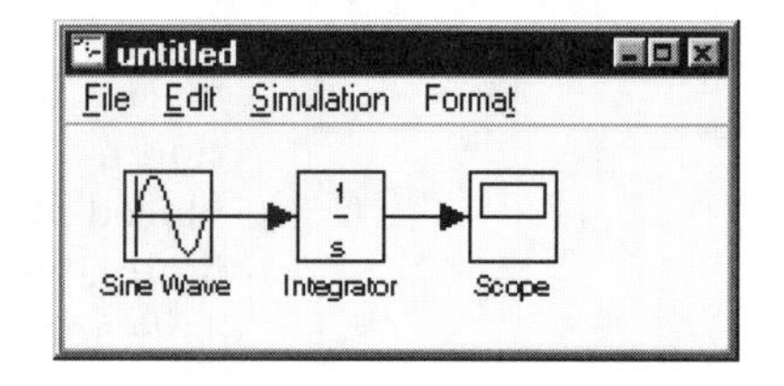

Double click on the Scope block, opening a Scope window as shown in Figure 2-6.

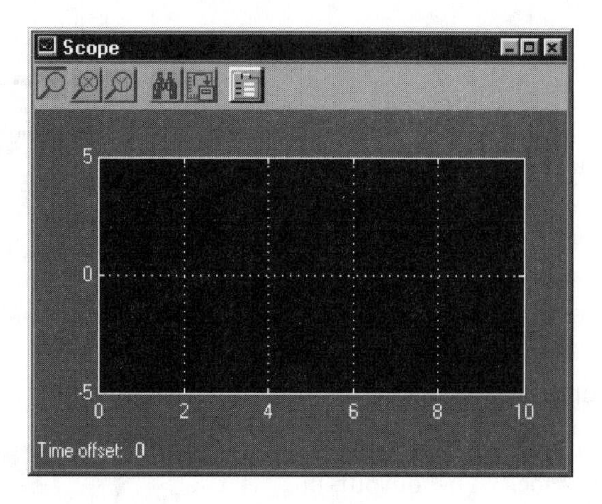

Figure 2-6 Scope window

Choose **Simulation:Start** from the menu bar in the model window (not the SIMULINK block library window). The simulation will execute, resulting in the scope display shown in Figure 2-7.

To verify that the plot shown in Figure 2-7 represents the solution to Equation (2-1), you can solve Equation (2-1) analytically, with the result $x(t) = 1 - \cos(t)$. Thus, the SIMULINK model solved the differential equation correctly.

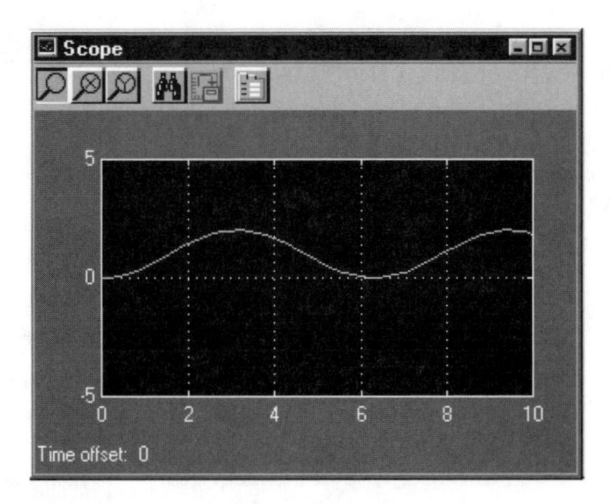

Figure 2-7 Scope display after executing the model

2.3 A More Complicated Model

So far, we've shown how to build a simple model. There are a number of additional model building skills that you'll need to acquire. In this section we use a model of a biological process to illustrate several additional skills: branching from signal lines, routing signal lines in segments, flipping blocks, configuring blocks, and configuring the simulation parameters.

Scheinerman [1] describes a simple model of bacteria growth in a jar. Assume that the bacteria are born at a rate proportional to the number of bacteria present, and that they die at a rate proportional to the square of the number of bacteria present. If x represents the number of bacteria present, then the bacteria are born at the rate

$$\text{birth rate} = bx \tag{2-2}$$

and they die at the rate

$$\text{death rate} = px^2 \tag{2-3}$$

The total rate of change of bacteria population is the difference between birth rate and death rate. This system can therefore be described with the differential equation

$$\dot{x} = bx - px^2 \tag{2-4}$$

Let's build a model of this dynamical system assuming that $b = 1/\text{hour}$ and $p = 0.5/\text{bacteria hour}$. Then we'll compute the number of bacteria in the jar after 1 hour, assuming that initially there are 100 bacteria present.

If you haven't closed SIMULINK yet, open a new model by choosing **File:New** from the SIMULINK block library window menu bar or clicking on the SIMULINK icon in the MATLAB window menu bar. Otherwise, open a new model window as discussed earlier.

This is a first-order system, so one integrator is required to solve the differential equation. The input to the integrator is $\dot{x}$, and the output is x. Open the Linear block library and drag the Integrator block to the position shown. Don't close the Linear block library yet.

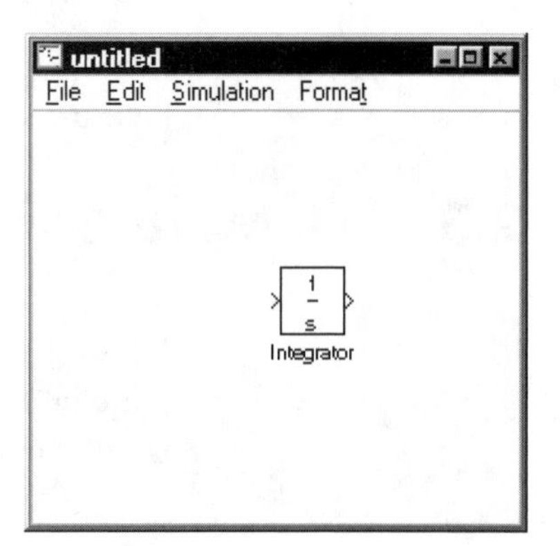

Drag two Gain blocks from the Linear block library and place them as shown. Notice that the name of the second Gain block is Gain1. SIMULINK requires each block to have a unique name. We'll discuss changing the name later.

Drag a Sum block from the Linear block library, and then close the Linear block library.

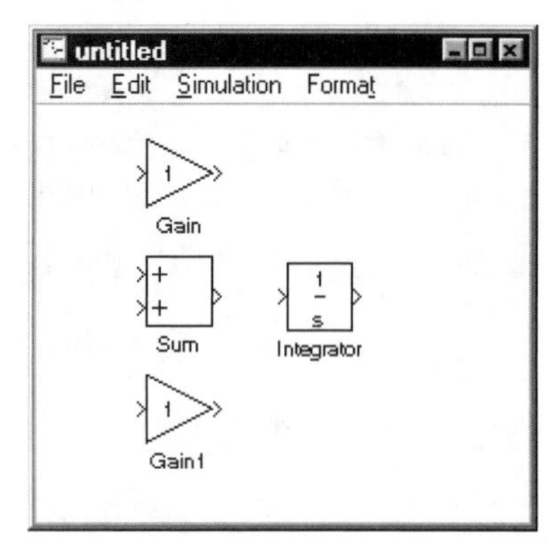

Open the Nonlinear block library and drag a Product block to the position shown. We will use the Product block to compute x^2.

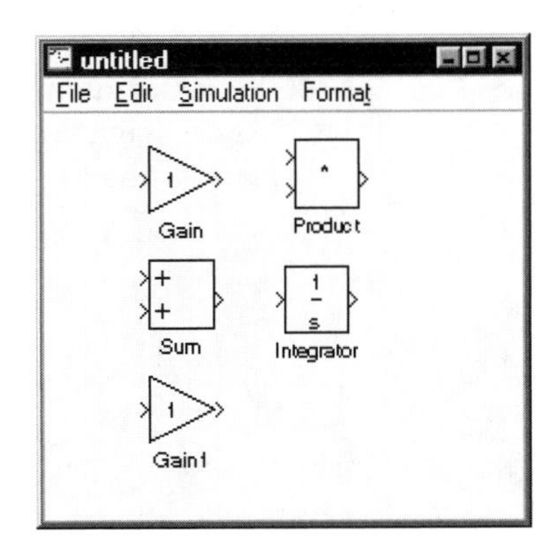

Open the Sinks block library and drag a Scope block to the model window as shown.

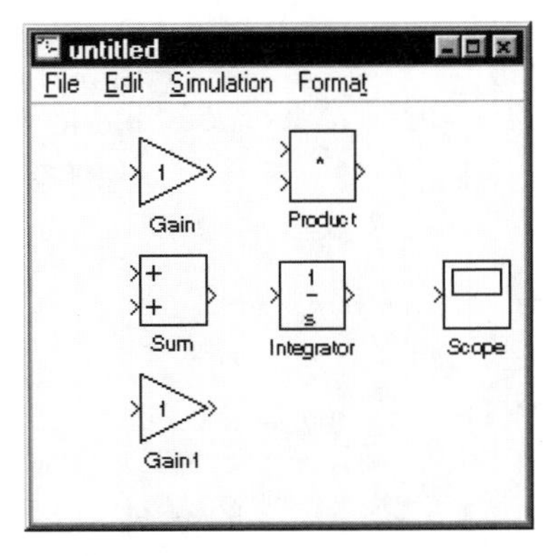

The default orientation of all the blocks is input ports on the left edge of the block and output ports on the right edge. The model will be much easier to read if we flip the Product block and the Gain blocks so that the input ports are on the right edge and the output ports are on the left edge. Starting with the Product block, click once on the block to select it. Notice the handles that appear at the four corners of the block, indicating that it is selected.

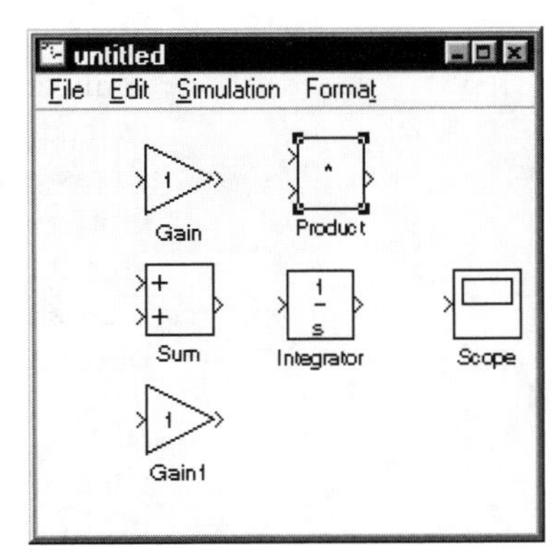

Choose **Format:Flip Block** from the model window menu bar. Now the inputs are on the right and the output is on the left. Repeat the flipping operation for each Gain block.

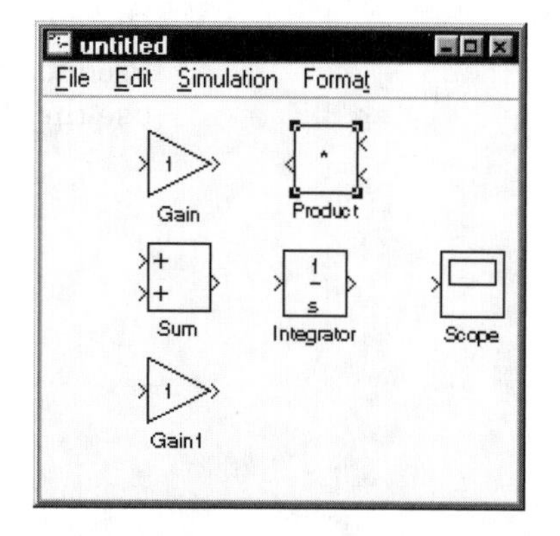

Draw a signal line from the output of the Sum block to the input of the Integrator block, and another from the output of the Integrator to the input of the Scope block.

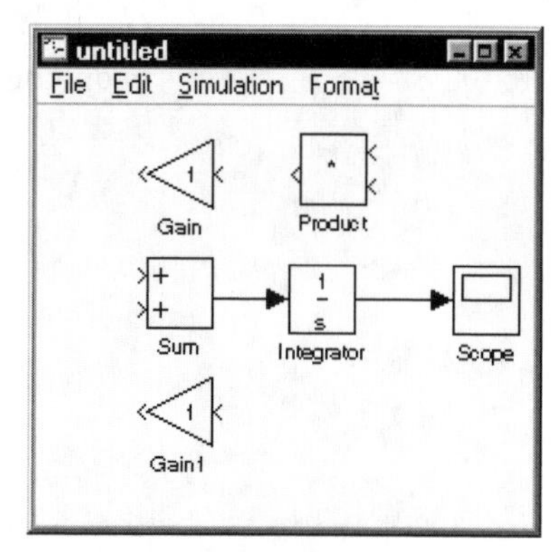

Next, we need to branch from the signal line connecting the Integrator and Scope blocks to feed the value of x to the lower Gain block. Press and hold the Ctrl key (Windows) or Option key (Macintosh) and click on the signal line. The cursor will change to a cross-hair shape. Continue to depress the mouse button and release the key.

If the mouse has two or three buttons, clicking and dragging using the right mouse button is equivalent.

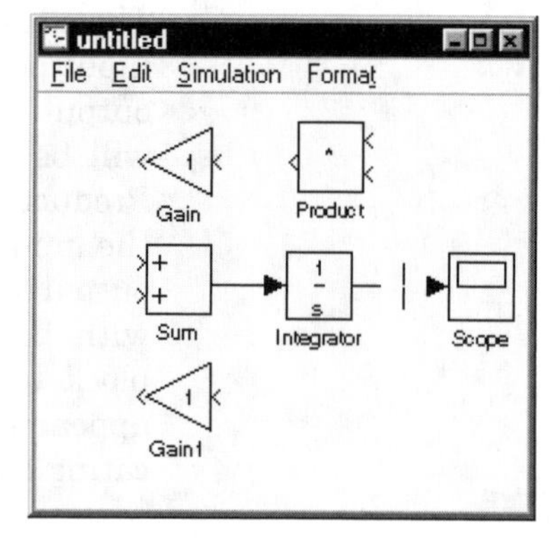

Drag directly to the input port of the Gain block. Notice that the signal line is dashed, and the cursor changes to a double cross-hair when it is on the input port of the Gain block. SIMULINK automatically routes the signal line using 90-degree bends.

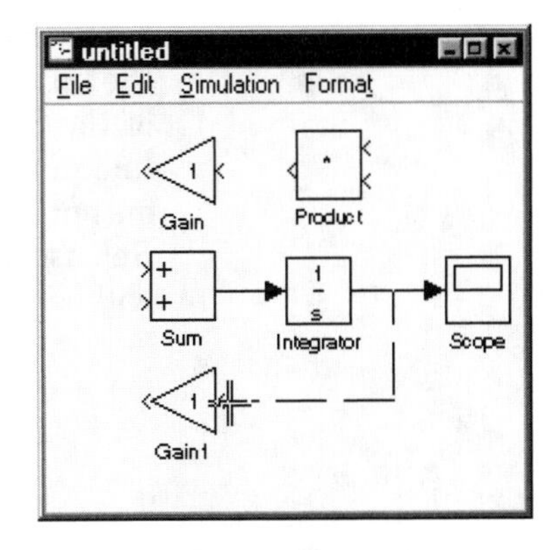

In a similar manner, branch from the signal line connecting the Integrator and Scope blocks to the top input port of the Product block.

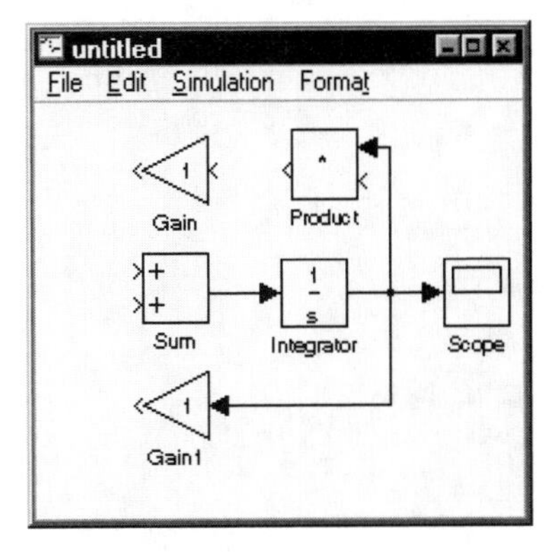

Branch from the signal line entering the upper input port of the Product block to the lower input port of the Product block. Thus, the output of the Product block is x^2. Connect the output of the Product block to the input of the upper Gain block.

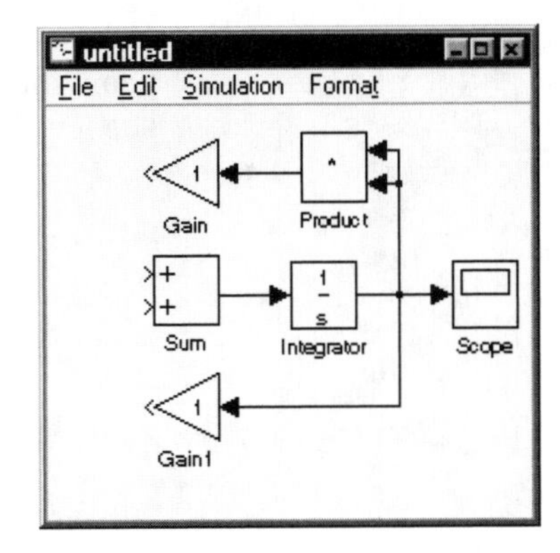

Draw the signal line from the output port of the upper Gain block to the upper input port of the Sum block in segments. To draw the line in segments, start by dragging from the output port to the location of the first bend. Release the mouse button. The signal line will be terminated with an open arrowhead.

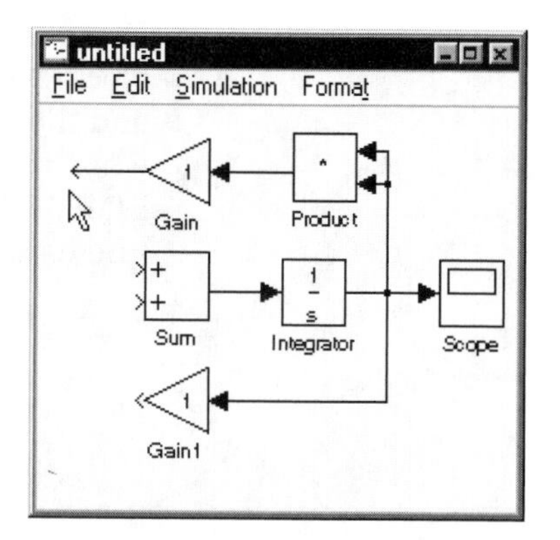

Next, drag from the open arrowhead to the upper input port of the Sum block.

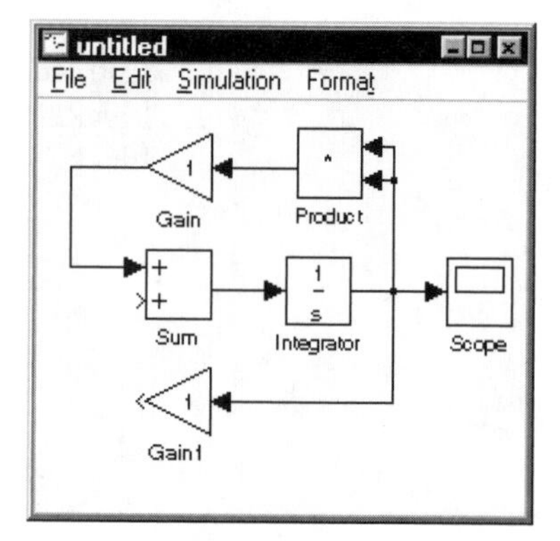

In a similar manner, draw the signal line from the output of the lower Gain block to the lower input of the Sum block.

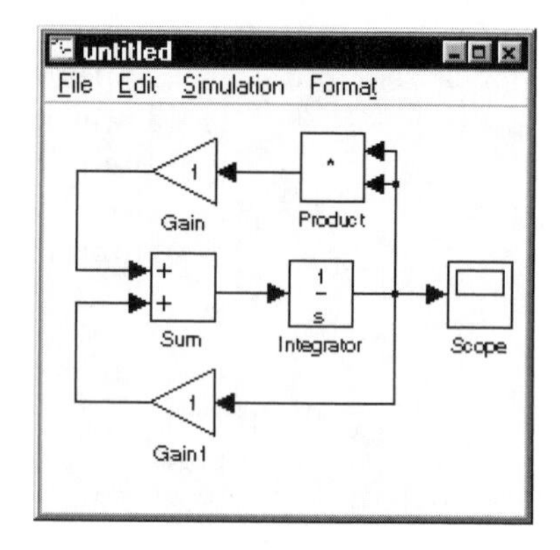

The model is now complete, but several of the blocks must be configured. Currently the value of gain for both Gain blocks is the default value of 1.0. The Sum block adds its two inputs instead of computing the difference. Finally, the initial value of the integrator output (the initial number of bacteria, x) must be set, as it defaults to 0.0. Start with the Gain blocks.

Double click on the upper Gain block. The Gain block dialog box will be displayed. Change the default value in field **Gain** to `0.5`. Click on **Close**.

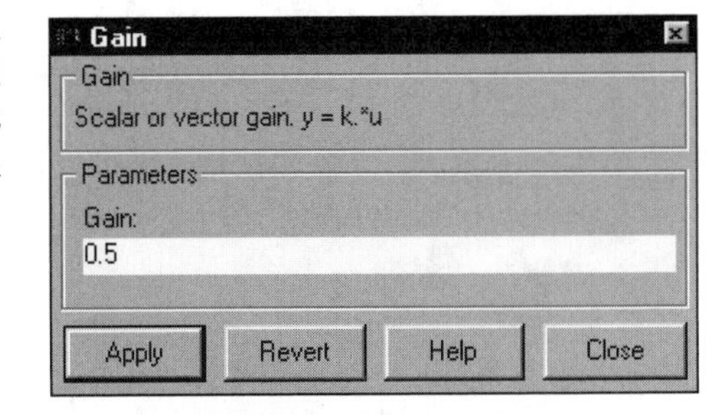

Notice that the value of gain on the block icon is now changed to 0.5.

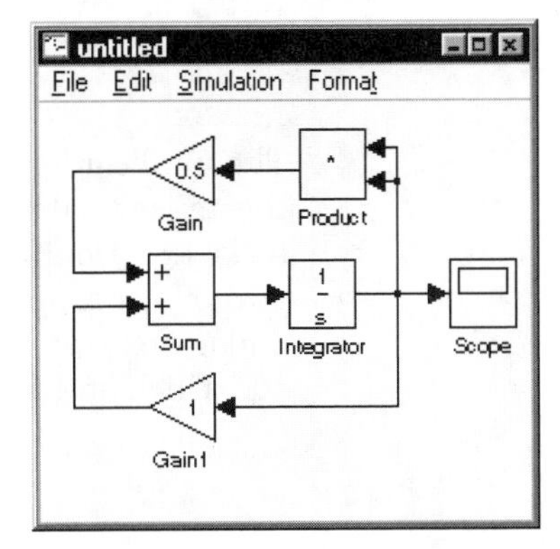

Double click on the Sum block, opening the Sum block dialog box. Set **List of signs** to −+ (a minus sign (–) followed by a plus sign (+)). Click **Close**.

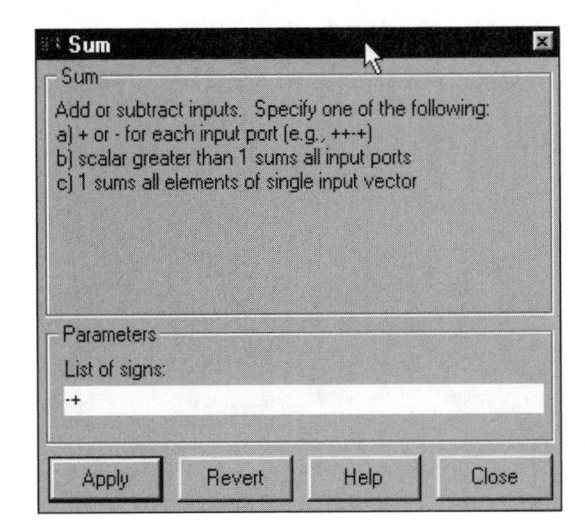

Now the Sum block is configured to compute the value of $\dot{x}$ according to Equation (2-4), after substituting in the values of b and p.

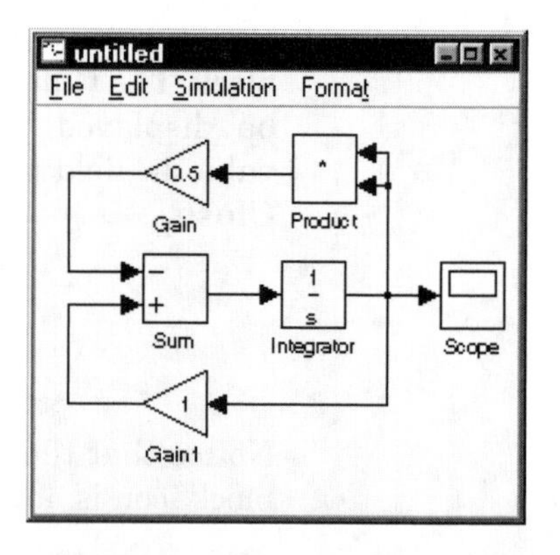

The final configuration task is to set the initial value of number of bacteria (x). Double click on the Integrator block, opening the Integrator dialog box. Set **Initial condition** to 100. Click on **Close**.

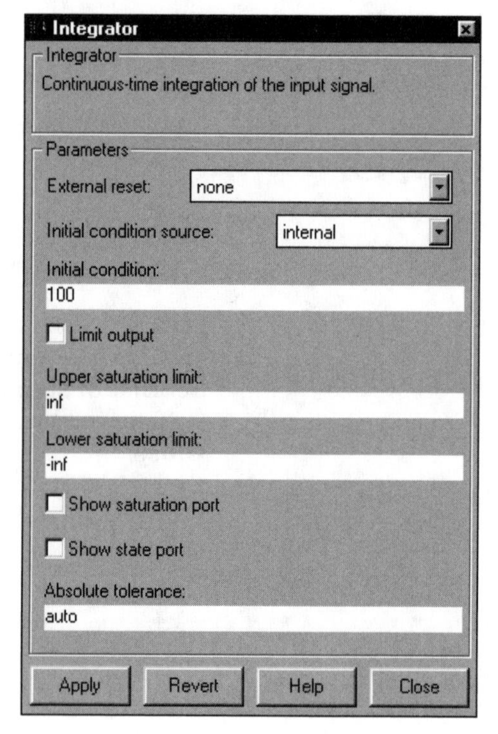

The simulation start time defaults to 0, and the stop time defaults to 10.0. To change the stop time to 1, open the Simulation parameters dialog box by choosing **Simulation:Parameters** from the model window menu bar. Set **Stop time** to 1, and then **Close**.

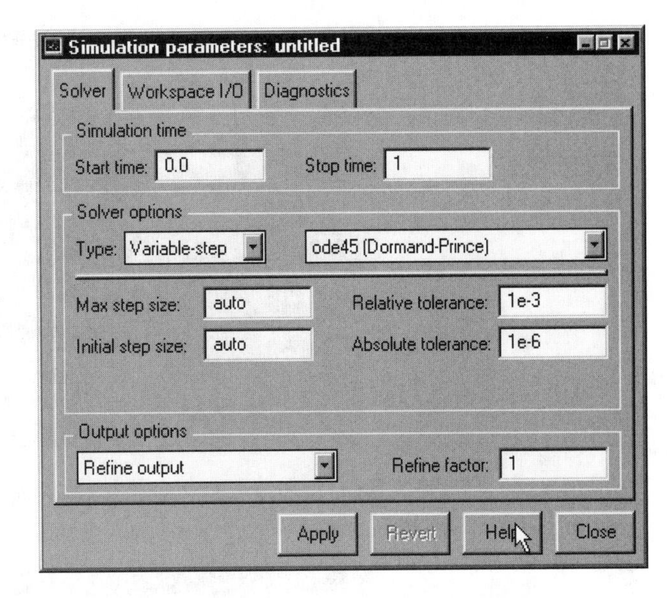

The model is now complete and ready to run. It's always a good idea to save a model before running it. To save the model, choose **File:Save** from the model window menu bar and enter a file name, say `examp_2`, without an extension. SIMULINK will save the model with the extension `.mdl` and change the model window name from `untitled` to the name you entered.

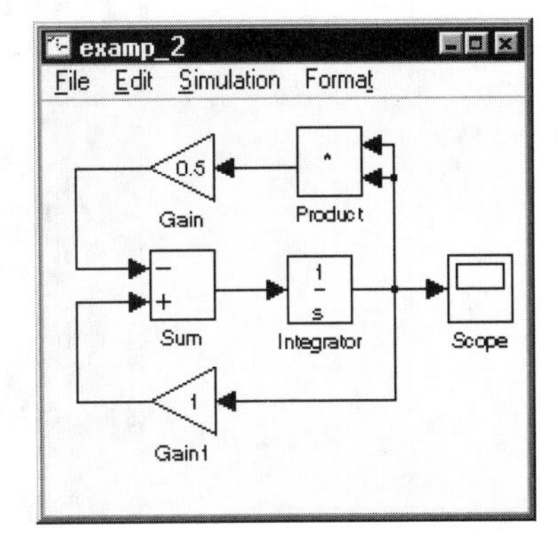

Open the Scope by double clicking on the Scope block. Then choose **Simulation:Start** to run the simulation. The Scope display will be as shown in Figure 2-8. Click on the autoscale button. The Scope will resize the scale to fit the entire range of values, as shown in Figure 2-9.

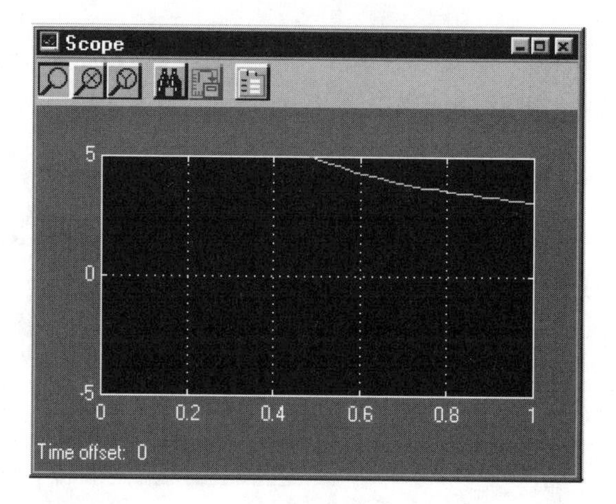

Figure 2-8 Scope display after running the bacteria growth model

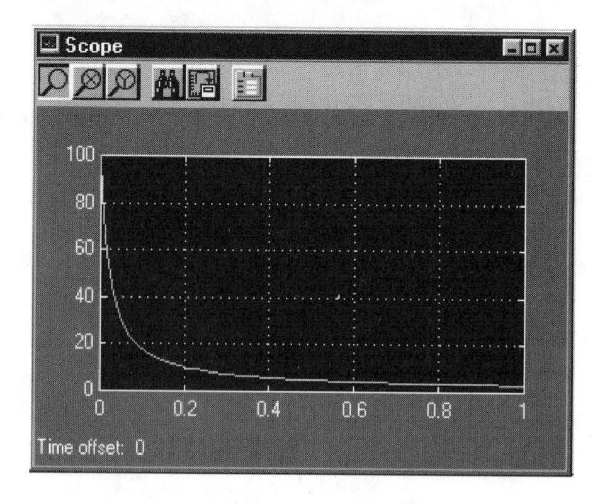

Figure 2-9 Scope display after rescaling

2.4 The SIMULINK Help System

SIMULINK includes an extensive on-line help system. The help files are designed to be viewed with a Web browser such as Netscape Navigator 3.0 or Microsoft Internet Explorer 3.0. Detailed on-line documentation for all of the blocks in the SIMULINK block library is available through the Block Browser, shown in Figure 2-10 Additionally, on-line help is available by clicking on the Help button in the dialog box for **Simulation:Parameters**.

2.4.1 Opening the Block Browser

The procedure for opening the Block Browser is as follows:

Double click on a block,

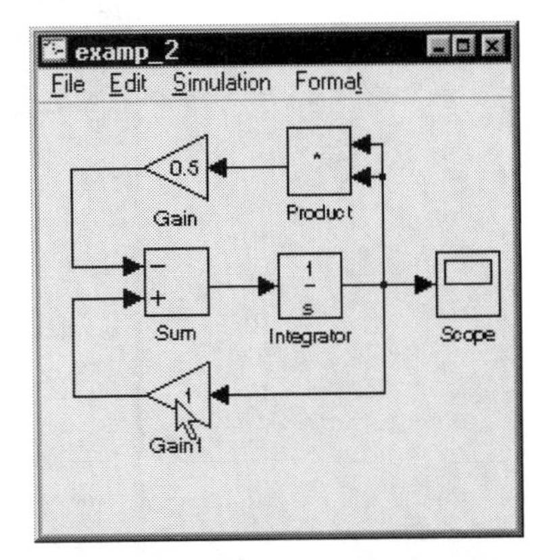

opening the block dialog box. Click on the **Help** button.

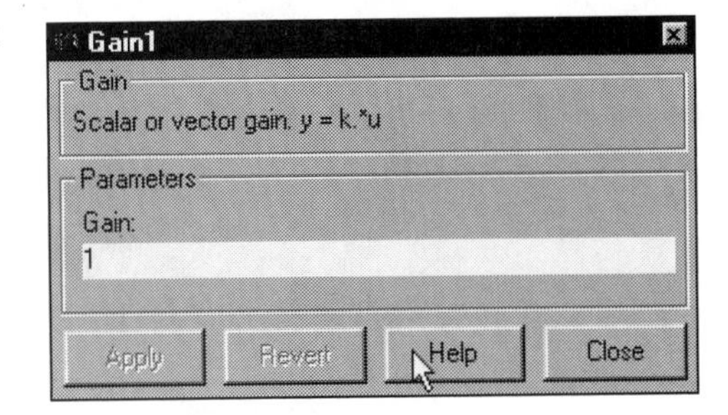

The Web browser (Netscape in Figure 2-10) will open to the Block Browser page for the block you selected.

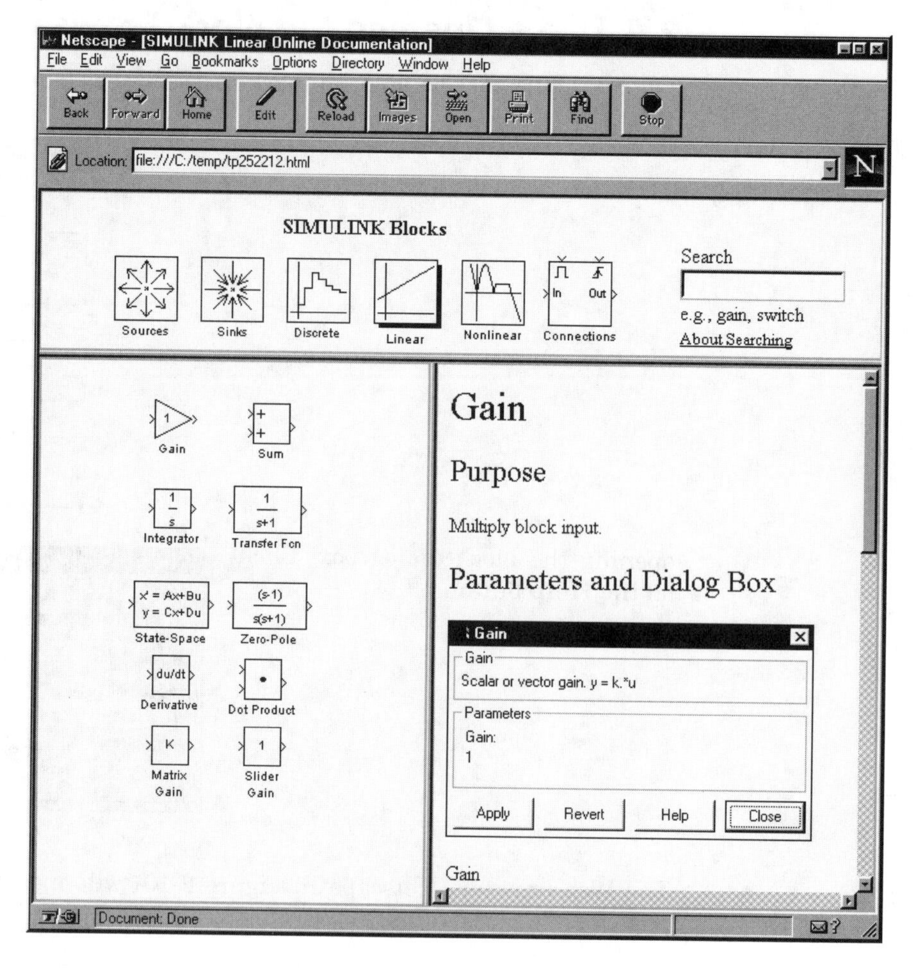

Figure 2-10 SIMULINK Block Browser

2.4.2 Block Browser Window

There are three frames to the Block Browser window. The top frame contains an icon for each block in the SIMULINK block library. The icon currently selected (the Linear block library in Figure 2-10) is highlighted with a drop shadow. Select a different block library by clicking on its icon in the top frame.

The lower left frame contains the icons for all blocks in the block library currently selected in the top frame. Click on a block in this frame to open the corresponding Block Browser page.

The top frame of the Block Browser also contains a field to request a search for a particular block. You can use this search capability to locate a Block Browser page quickly if you're not sure which library a block is in.

2.5 Summary

In this chapter we have presented the basic steps for building and executing SIMULINK models. We have also described the SIMULINK help system. Using the procedures discussed in this chapter, it is possible to model a wide variety of dynamical systems. Before proceeding further, you may find it beneficial to build a few simple models to gain proficiency in the skills discussed thus far. You may also find it interesting to explore some of the SIMULINK demonstrations in the Demos block library.

2.6 Reference

1 Scheinerman, Edward C., *Invitation to Dynamical Systems*. Upper Saddle River, N.J.: Prentice Hall, 1996, pp 22–24.

3

Model Building

In this chapter, we will explain the mechanics of model building in detail. The procedures discussed here will enable you to build models that are easy to interpret. You will also learn how to select and configure a differential equation solver and how to print a SIMULINK model and embed a model in a word processing document.

3.1 Introduction

In Chapter 2 we discussed the basics of building and running a SIMULINK model. Using the procedures discussed there and the on-line Block Browser, you can build extremely complex models. However, as models get more complex, they become more difficult to interpret. The procedures discussed in this chapter will enable you to make your models easier to understand. First, careful arrangement of blocks and signal lines can make the relationships easier to follow. Next, naming blocks and signal lines and adding annotations to the model can make the purpose of the model elements easier to understand.

We will also describe the **Simulation:Parameters** dialog box, which provides extensive options for selecting and configuring the differential equation solver used to perform model simulation. In addition to selecting a solver most appropriate to your problem, you can control the spacing of output points and the generation of error and warning messages, and even send internal simulation data to the MATLAB workspace.

Finally, we will explain how to print your models. You can either print directly to a printer or embed an image of the model in a word processing document.

Our purpose in this chapter is to explain the mechanics of model building: manipulating blocks, drawing and editing signal lines, annotating the model, and so on. Once you have mastered these mechanics, you will be ready to begin building models using the procedures covered in the subsequent chapters.

3.1.1 Elements of a model

A SIMULINK model consists of three types of elements: *sources*, the *system* being modeled, and *sinks*. Figure 3-1 illustrates the relationship among the three elements. The central element, the system, is the SIMULINK representation of a block diagram of the dynamical system being modeled. The sources are the inputs to the dynamical system. Sources include constants, function generators such as sine waves and step functions, and custom signals you create in MATLAB. Source blocks are found in the Sources block library. The output of the system is received by sinks. Examples of sinks are graphs, oscilloscopes, and output files. Sink blocks are found in the Sinks block library.

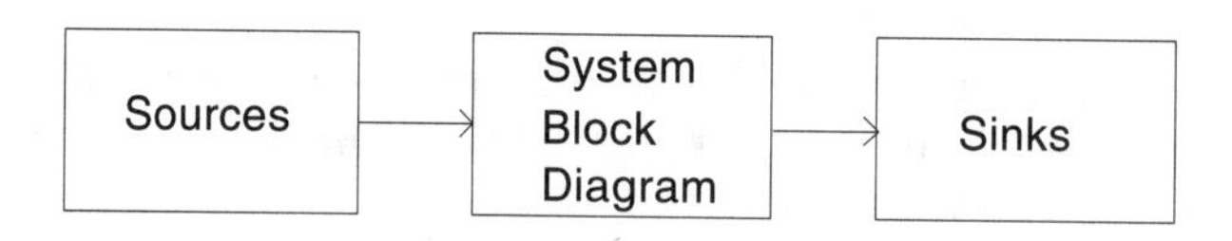

Figure 3-1 Elements of a SIMULINK model

Frequently, SIMULINK models lack one or more of these three elements. For example, you might wish to model the unforced behavior of a system initially displaced from its equilibrium state. Such a model would have no inputs, but it would have system blocks (Gain blocks, Integrators, etc.) and probably sinks. It is also possible to build a model that has sources and sinks, but no system blocks. Suppose you need a special signal composed of the sum of several functions. You could easily generate the signal using SIMULINK source blocks and send the signal to the MATLAB workspace or to a disk file.

3.2 Opening a Model

In Chapter 2 we discussed creating a new model by choosing **File:New** from the SIMULINK menu bar or by clicking on the SIMULINK icon in the MATLAB tool bar. We also discussed saving a model using **File:Save**. To use an existing model, choose **File:Open** from the menu bar of the SIMULINK window and select the file.

The default directory will be the current default directory in the MATLAB session from which you started SIMULINK. If you change to a different directory using the SIMULINK file menu, the default directory will not change. It will remain the same as the current directory in the MATLAB session, so if you wish to open another model in the same directory as the model you previously opened and that model was not in the directory currently active in the MATLAB session, you will have to navigate once again to the appropriate directory. Therefore, it is frequently convenient to change to the directory containing the SIMULINK model in the MATLAB session (using the `cd` command) and then open the model using the SIMULINK **File:Open** menu choice.

Although it is not necessary, we recommend that you set up a directory for your SIMULINK models separate from the MATLAB directory structure. This will make it easier to keep track of your files. It will also eliminate the possibility of overwriting or deleting your files if you upgrade MATLAB and SIMULINK in the future.

An alternate method to open an existing model is to change to the appropriate directory in the MATLAB session and then type the name of the model as a MATLAB command. For example, if the model is named `examp.mdl`, enter the command `examp` at the MATLAB prompt.

3.3 Manipulating Blocks

In Chapter 2 you learned how to drag a block from a block library to a model window and to flip a block. In addition to those basic operations, you can resize, rotate, copy, and rename blocks. In this section we will discuss these and several other block manipulation operations. To prepare to practice the operations illustrated here, start SIMULINK, or if SIMULINK is already started, open a new model window.

3.3.1 Resizing a Block

Open a block library, and drag a block to the model window. Here, we're using an Integrator block from the Linear block library.

To resize a block, first select the block, causing the handles to appear.

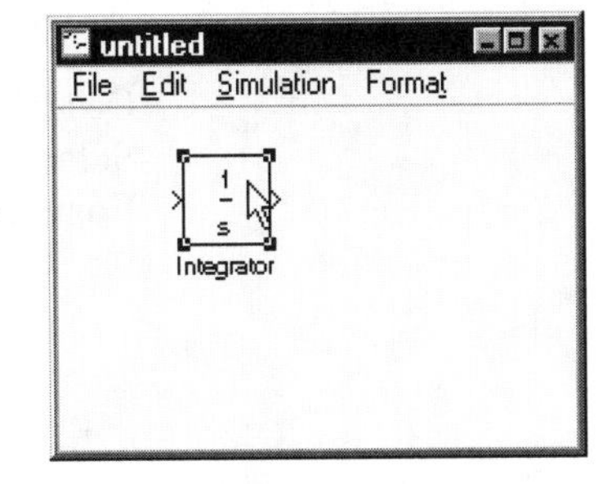

Click on the desired handle and, continuing to depress the mouse button,

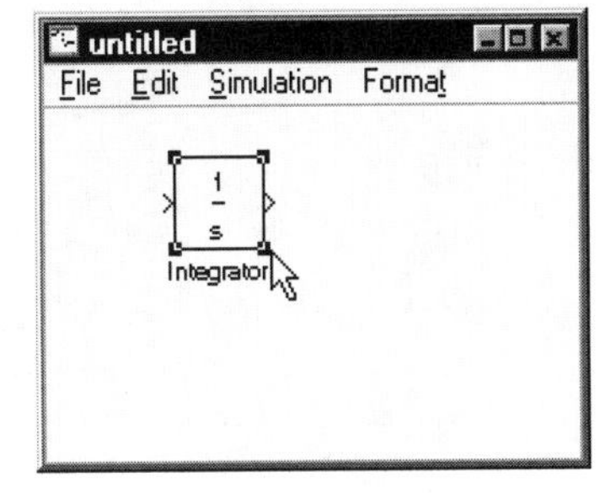

drag the handle to resize the block. Notice that the cursor changes shape, confirming that you have "grasped" the resize handle.

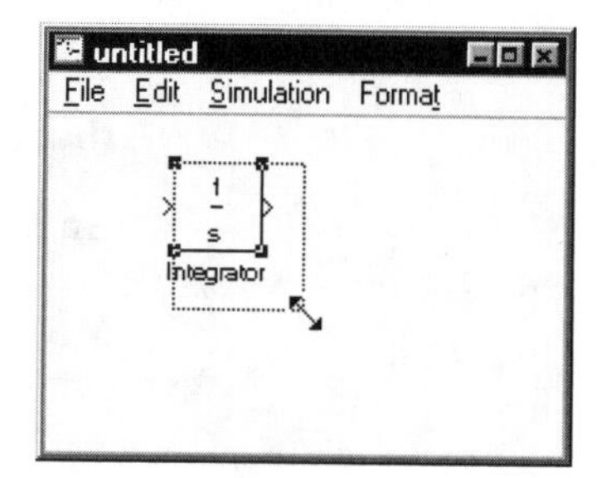

Release the mouse button.

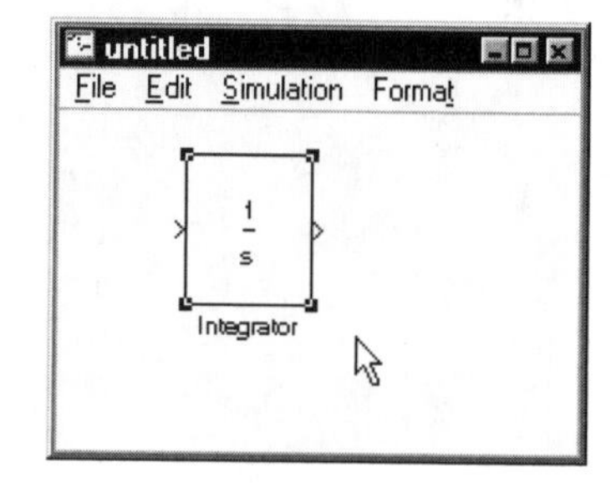

3.3.2 Rotating a Block

Occasionally, you will want to rotate a block. Select the block and then choose **Format:Rotate**.

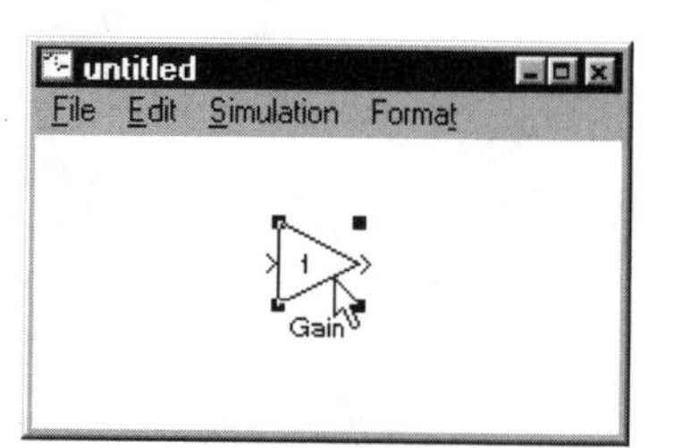

The block will rotate 90 degrees clockwise.

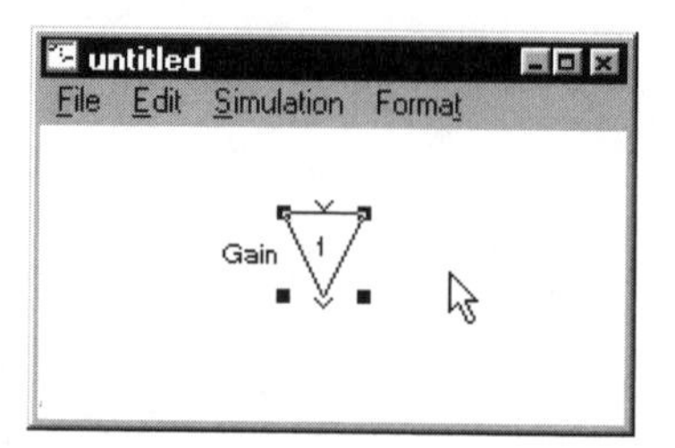

3.3.3 Copying a Block Within a Model

You will frequently want to copy a block from within a model. For example, after you have resized a block, you will probably want all similar blocks to appear identical. Rather than attempting to resize each block, you can copy the block you resized.

To copy a block within a model, depress and hold the Ctrl key (Windows) or Option key (Macintosh) and then click on the block.

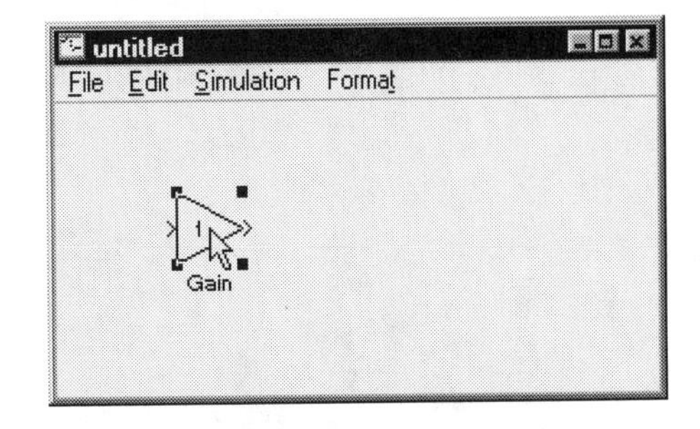

Drag the copy to the desired position.

If you have a two (or three) button mouse, dragging using the right mouse button is equivalent.

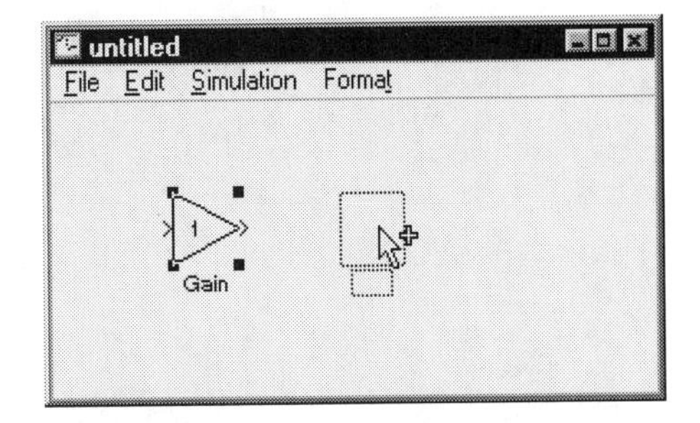

Release the mouse button, completing the copy operation.

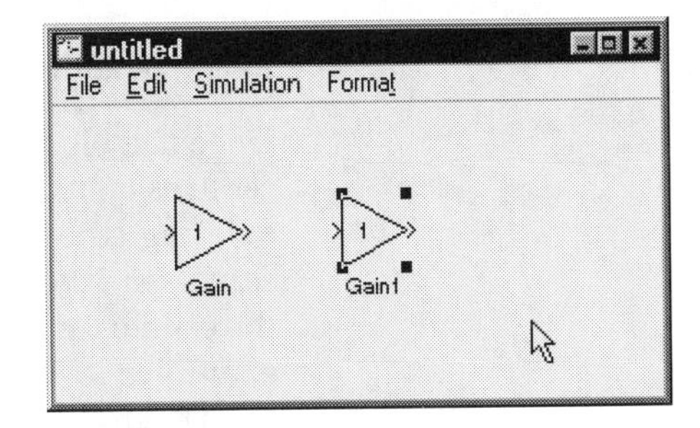

An alternative is to click on the block and select **Edit**:**Copy** (or press Ctrl-C in Windows, ⌘-C on Macintosh) and then **Edit**:**Paste** (or Ctrl-V in Windows, ⌘-V on Macintosh).

3.3.4 Deleting Blocks

To delete a block, select the block and then press the Delete key. An alternative is to select the block and then choose **Edit:Clear** from the model window menu bar. Choose **Edit:Cut** to delete the block and save it on the Clipboard.

3.3.5 Selecting Multiple Blocks

You can select multiple blocks and move, copy, or delete them as a group. There are two ways to select multiple blocks. The first is to depress and hold the Shift key while clicking on each block in the group. The other method is to use a bounding box as follows:

Click and hold the mouse button outside the blocks.

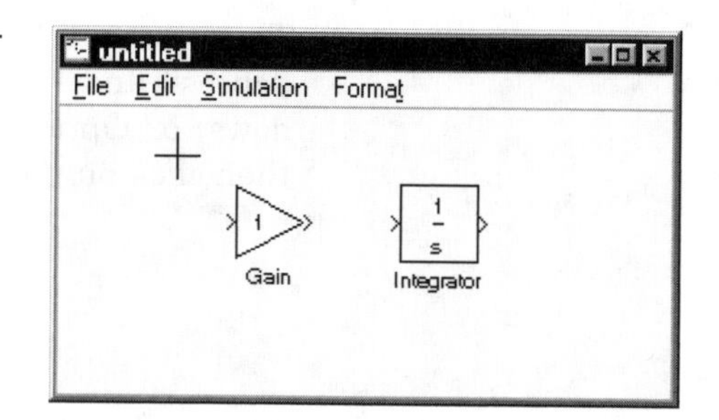

Drag the bounding box that appears to enclose all the desired blocks.

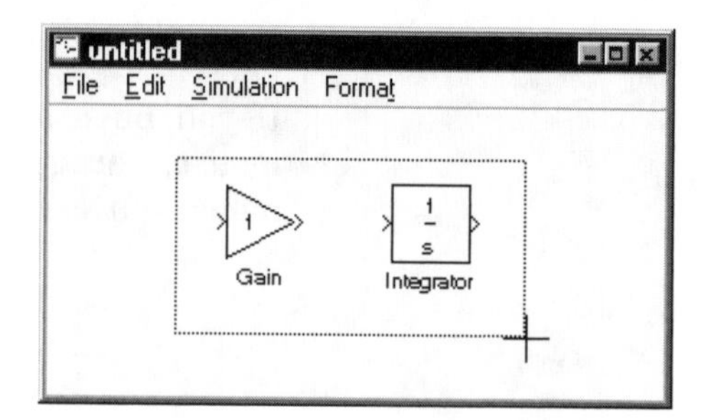

Release the mouse button. All blocks in the bounding box are now selected.

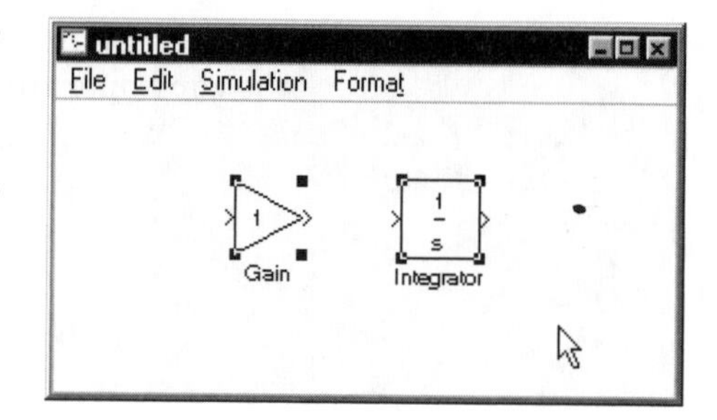

Once the blocks are selected, the procedures for moving, copying, or deleting the entire group are the same as the corresponding procedures for a single block. So, for example, to delete the group, press the Delete key or choose **Edit:Cut** or **Edit:Clear** from the model window menu bar.

3.3.6 Changing a Block Label

SIMULINK supplies a default label for each block as you place the blocks in the model window. For example, the first gain block will have the default name "Gain", the second gain block will be labeled "Gain1", and so on. Change the block label as follows:

Click on the block label. An editing cursor will appear. You can position the cursor anywhere in the label by clicking, and you can move the cursor using the cursor movement keys.

To replace a label, select it and then double click on it. The label will be highlighted as shown here.

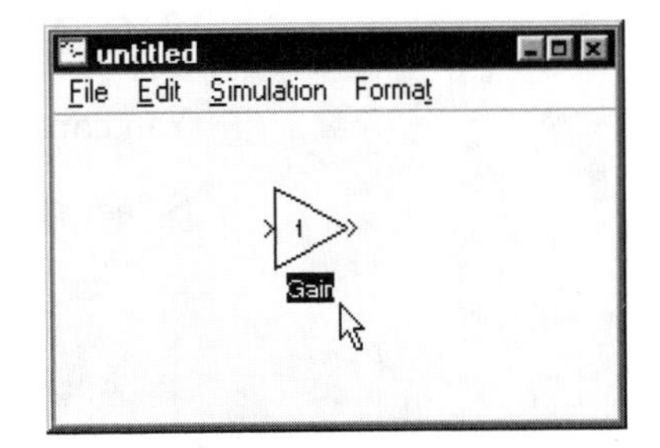

Enter the new label.

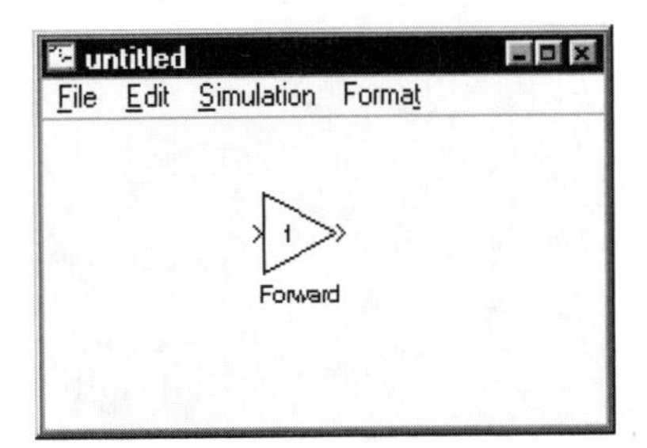

To create a multiple line label, press Return at the end of each line.

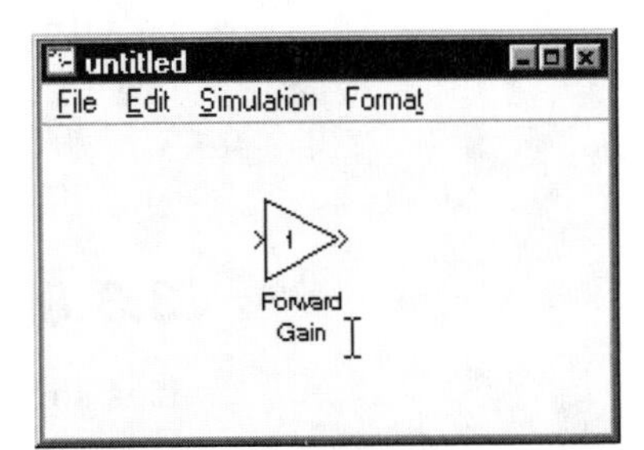

Click away from the block to accept the label.

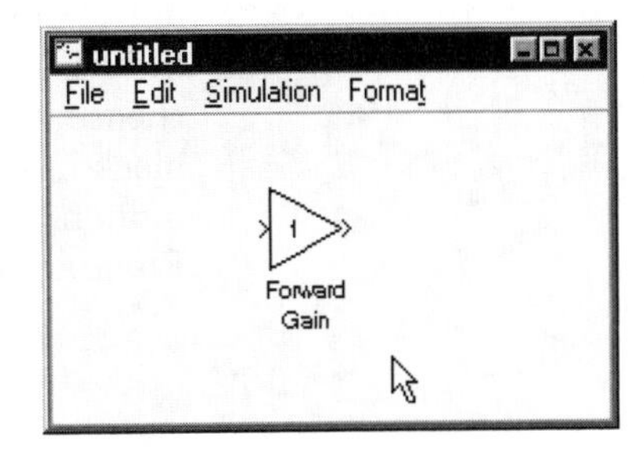

Each block must have a unique name of at least one character.

3.3.7 Changing Label Location

You can move the block label from below to above the block as follows:

Select the block.

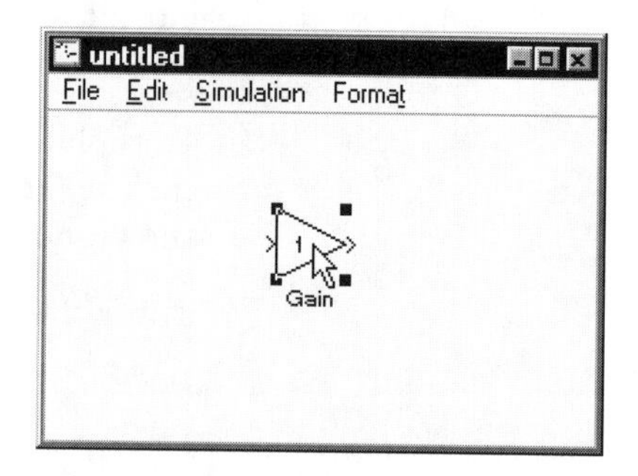

Choose **Format:Flip Name**.

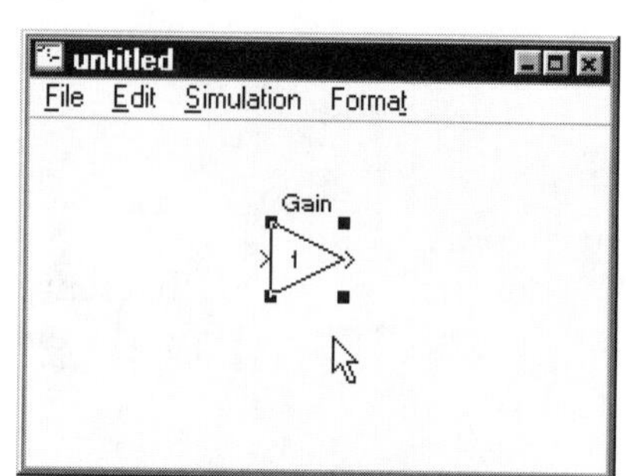

An alternative is to click on the name and drag it to the desired position. If the block is rotated 90 degrees, the block label will be either to the left or the right of the block.

3.3.8 Hiding a Label

It is sometimes desirable to hide the name of a block. For example, since the shape of a Gain block uniquely identifies the purpose of the block and the value of the gain is displayed on the block, a cluttered model might be improved by hiding the names of the Gain blocks. To hide the name, select the block and choose **Format:Hide Name.** Note that this does not change the name of the block in any way—it simply makes the name invisible. If the name of a block is hidden, when the block is selected **Format:Hide Name** is replaced by **Format:Show Name** on the **Format** pull-down menu.

3.3.9 Adding a Drop Shadow

If you want to call special attention to a block, you can apply a drop shadow as follows:

Select the block.

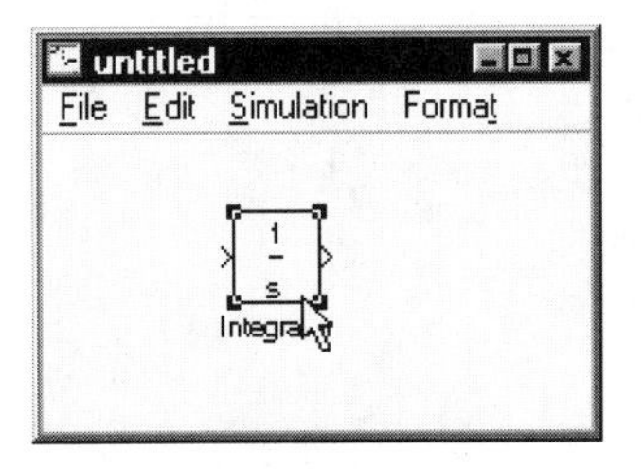

Choose **Format:Show Drop Shadow** from the model window menu bar. If a block is configured with a drop shadow, the menu choice will change to **Format:Hide Drop Shadow** for that block, permitting you to remove the drop shadow if desired.

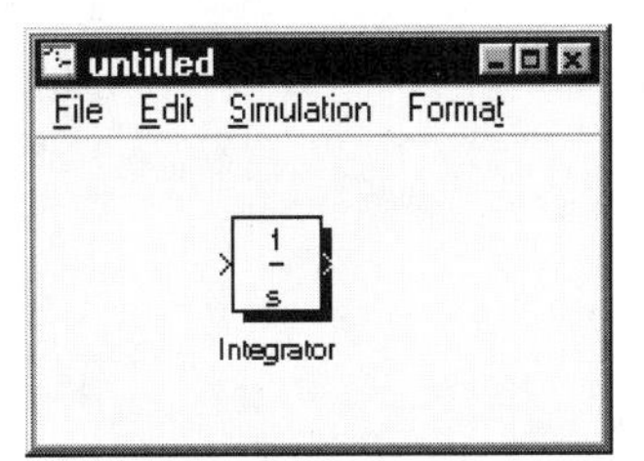

3.3.10 Configuring Blocks

In Chapter 2 we discussed setting block parameters. In the example in Chapter 2, the numeric parameters were set to constants. However, configuration parameters do not have to be constants. They can be any valid MATLAB expression and may use variables that will be defined in the MATLAB workspace when the model is executed. This is a very useful capability. Later we'll discuss executing a SIMULINK model from a MATLAB script. By making certain parameters variables, the parameters can be changed by the script.

3.4 Signal Lines

In Chapter 2 we discussed drawing signal lines using the automatic routing capability and drawing signal lines in segments. In this section we'll discuss drawing signal lines in an arbitrary direction and editing signal lines.

3.4.1 Drawing Signal Lines at an Angle

Recall that if you drag directly from an output port of one block to an input port of another block, SIMULINK will automatically route the signal lines in horizontal and vertical segments. If you override the automatic routing and draw a signal line in segments, the segments will still be horizontal and vertical. You can override this feature and draw signal lines at any arbitrary angle as follows:

Depress and hold the Shift key and then click and hold on the output port.

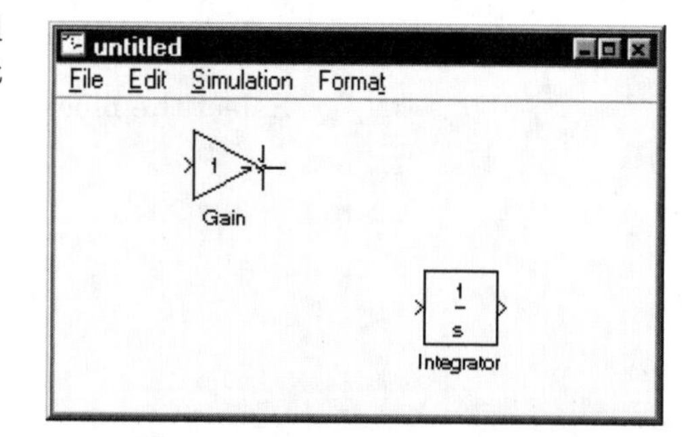

Drag the end of the signal line to the desired point.

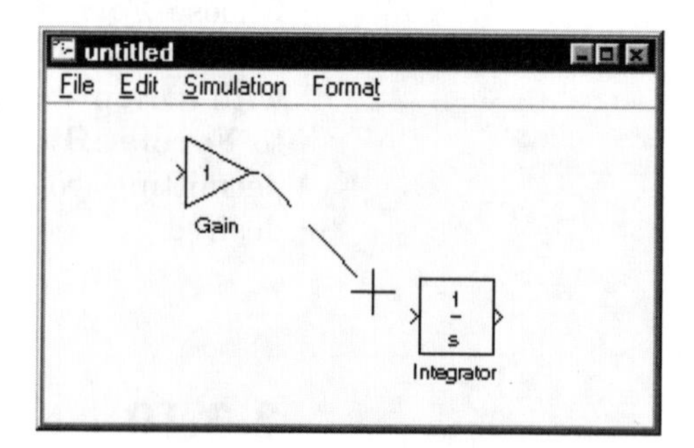

Release the mouse button and Shift key.

To draw in segments, release the mouse button momentarily at the end of each segment and then continue drawing the next segment, all the while continuing to depress the Shift key.

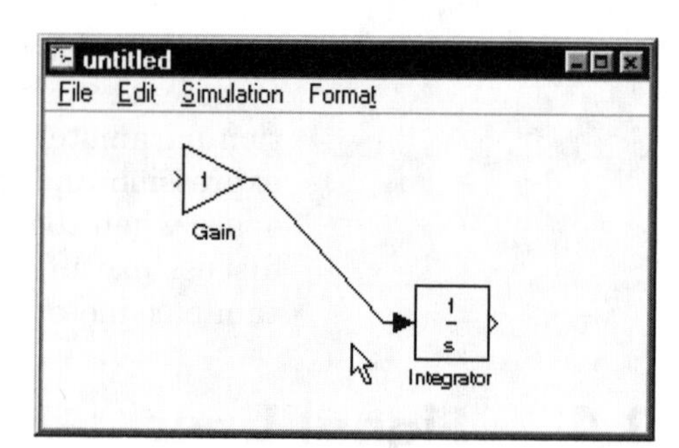

3.4.2 Moving a Segment

To move a line segment, click on the segment.

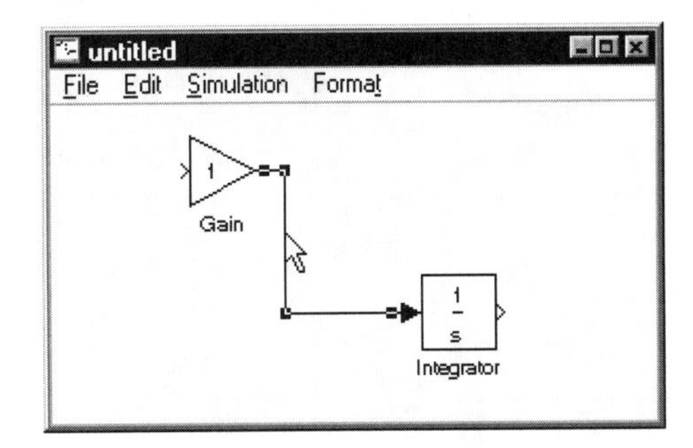

The cursor will change shape.

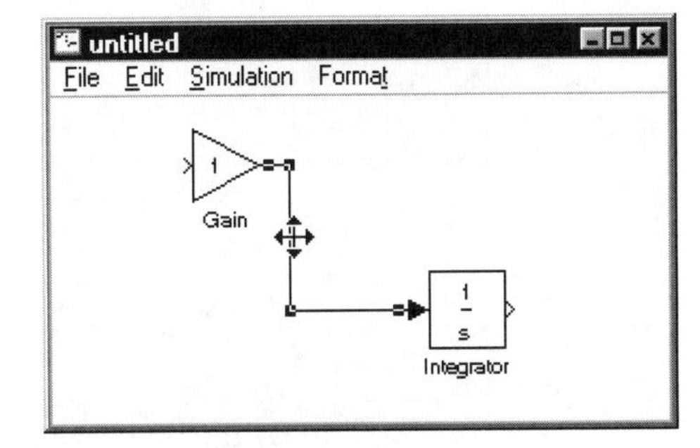

Holding the mouse button depressed, drag the segment to the desired location.

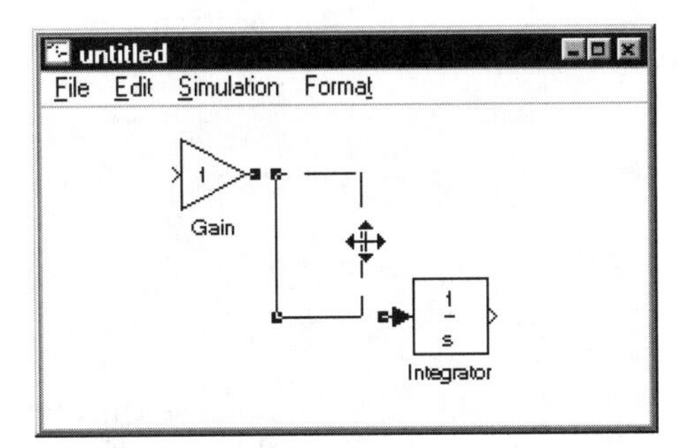

Release the mouse button.

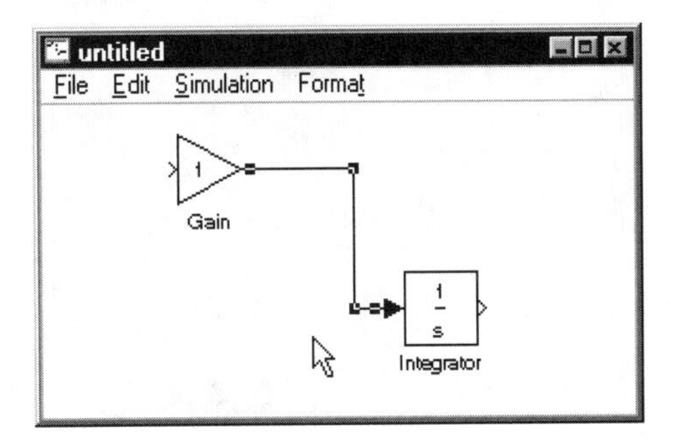

3.4.3 Moving a Vertex

To move a vertex, start by clicking on the vertex.

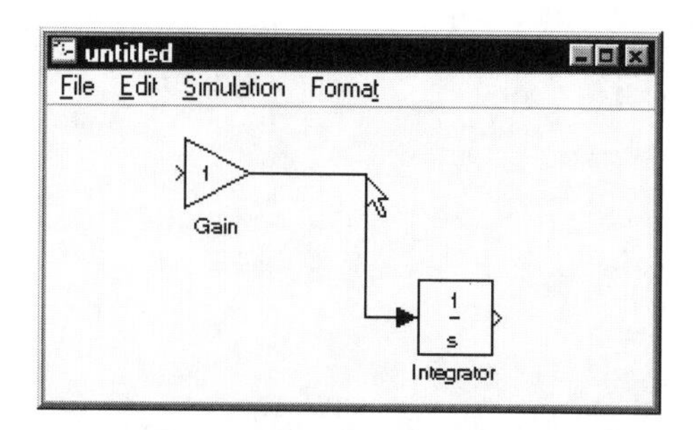

The cursor will change shape to a circle.

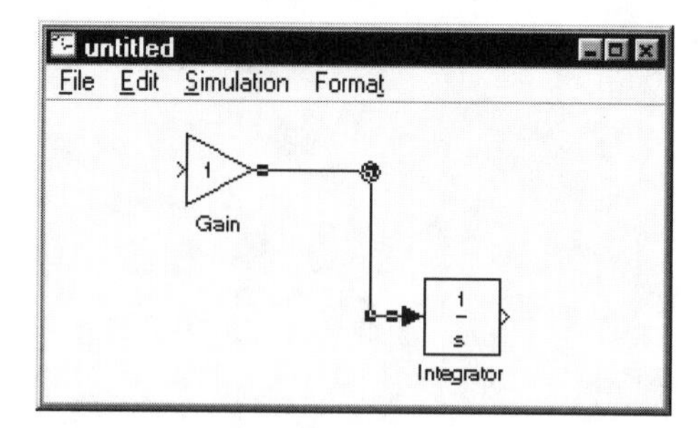

Drag the vertex to the desired location.

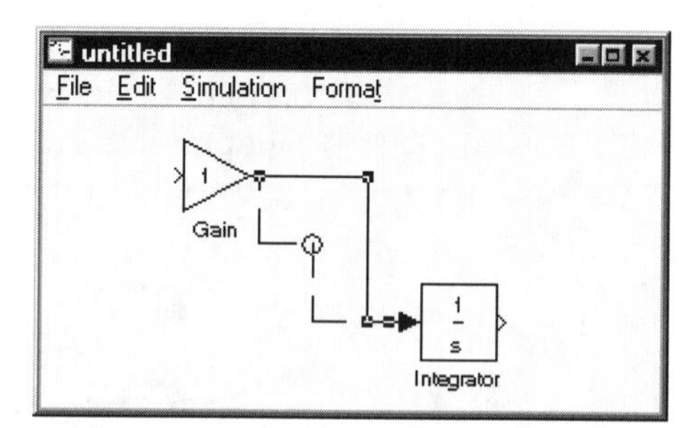

Release the mouse button to complete the operation.

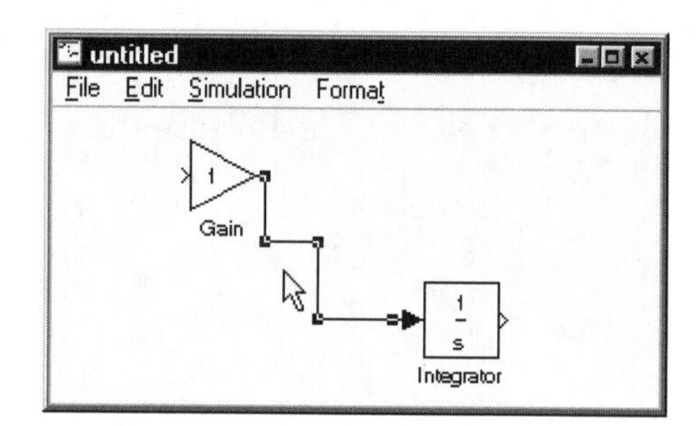

3.4.4 Deleting a Signal Line

To delete a signal line, select the signal line by clicking on it. Press the Delete key, or choose **Edit:Clear** or **Edit:Cut** from the model window menu bar.

3.4.5 Splitting a Signal Line

To split a line, first select the line. Depress the Shift key and then click on the line at the point at which you wish to split it, keeping the mouse button depressed.

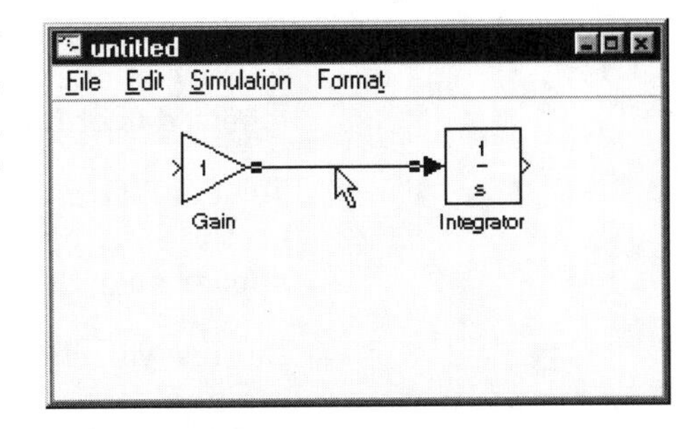

The cursor will change shape to a circle and the segment will split into two segments.

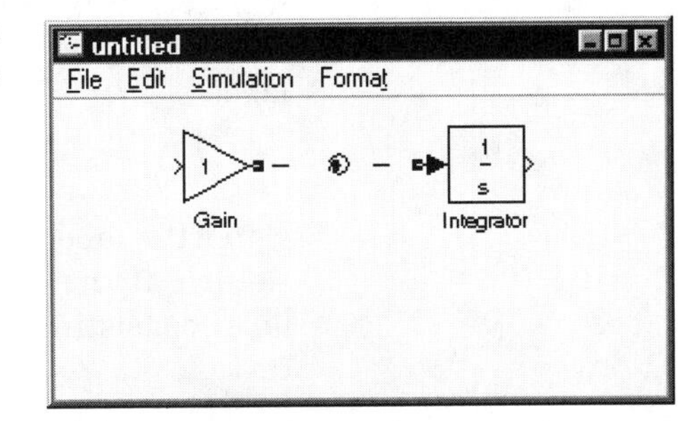

Drag the new vertex to the desired location.

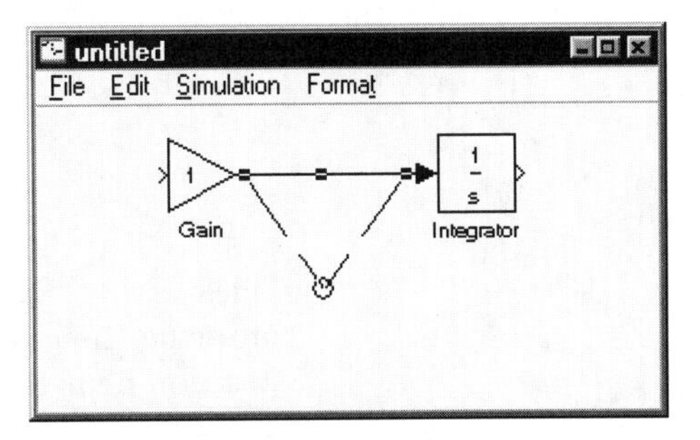

Release the mouse button to complete the process.

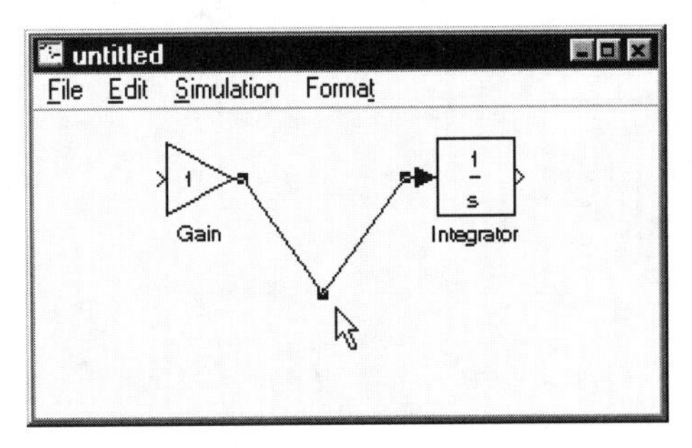

3.4.6 Labeling a Signal Line

Each signal line may have a label. The label can be positioned at the middle or either end of the signal line, and on either side. Signal line labels, unlike block labels, do not have to be unique.

To add a label to a signal line, double click on the line, causing an editing cursor to appear near the line.

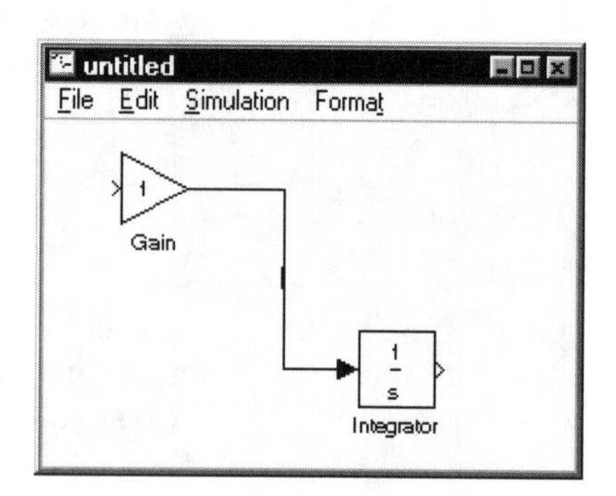

Be sure you click on the line, not just near the line. Double clicking away from a line will result in an annotation (to be discussed shortly) rather than a signal line label.

Enter the label. As with block labels, press Return at the end of each line to enter a label consisting of more than one line of text.

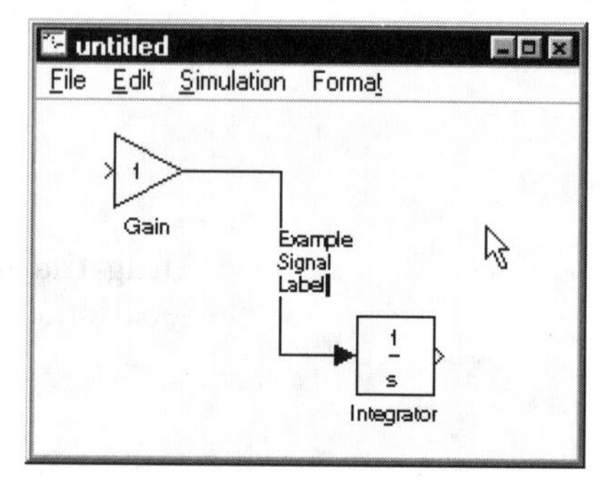

Click away from the line to complete entering the label. The label will snap to a position near the center of the line.

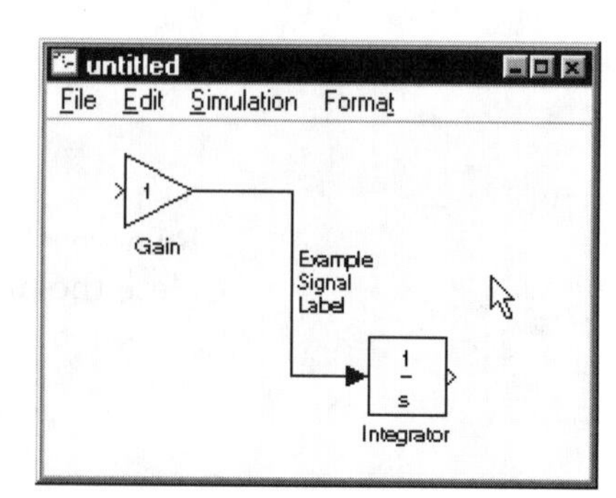

3.4.7 Moving or Copying a Signal Line Label

You can move a signal line label to either end or the middle of the signal line.

To move a signal line label, click on the label.

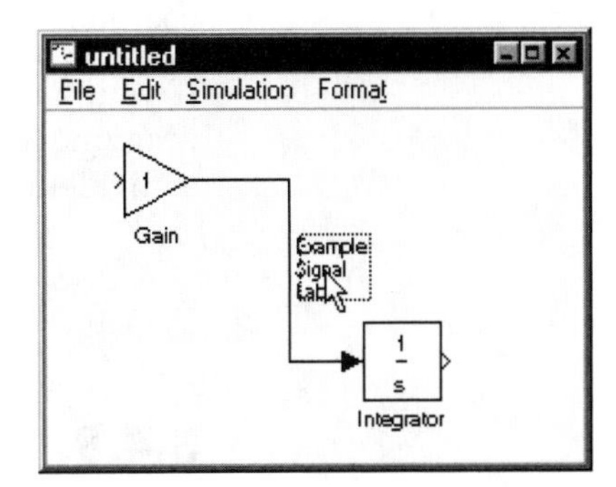

Drag the label to the desired location near the signal line.

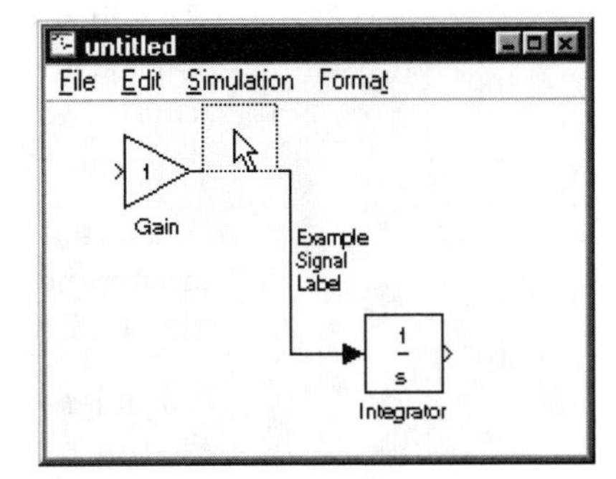

Release the mouse button. The label will snap into position.

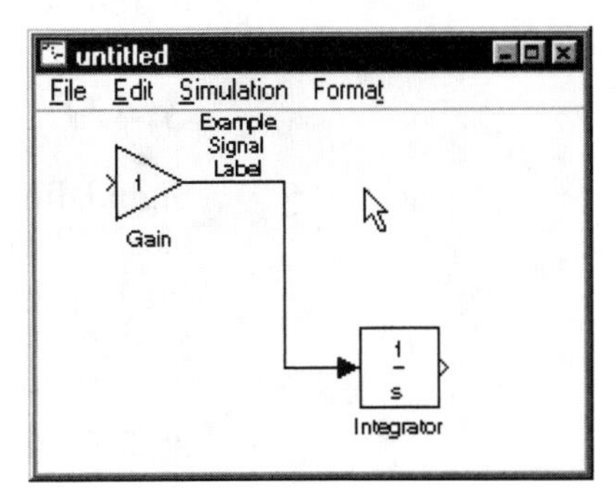

To copy a signal line label, depress the Ctrl key (Windows) or Option key (Macintosh) while dragging the label to the new position.

If your mouse has two or three buttons, dragging with the right mouse button is equivalent.

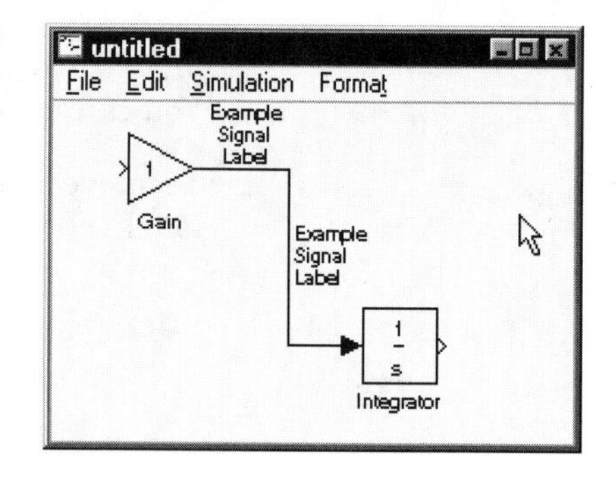

3.4.8 Editing a Signal Line Label

To edit a signal line label, click on the label. An editing cursor will appear. You can position the cursor anywhere in the label by clicking, or you can move the cursor with the cursor movement keys. Click away from the label when you're finished editing it. All occurrences of the label will be changed.

If there are multiple occurrences of a signal line label, you can delete a single occurrence. Depress the Shift key and then click on the occurrence you wish to delete. Press the Delete key to complete the process.

To delete all occurrences of a signal line label, delete all characters in one instance of the label. When you click away from the label, all occurrences will be removed.

3.4.9 Signal Label Propagation

Signal line labels can propagate through several blocks in the Connections block library. Among these are the Mux and Demux, Goto and From, and Inport and Outport blocks. Signal line propagation provides an accurate means to determine the exact content of a signal line. This can make a model easier to understand and can also be useful in debugging. The process is illustrated here.

We have a model in which two scalar signals produced by Constant blocks are combined by a Mux block to form a vector signal. The Demux block splits the vector signal into two scalar signals that are displayed by Display blocks. Notice that the model is shown after running the simulation (the Display blocks show the signal values).

First, label the source signal lines. Here the outputs of the constant blocks `KA` and `KB` are labeled `A` and `B`.

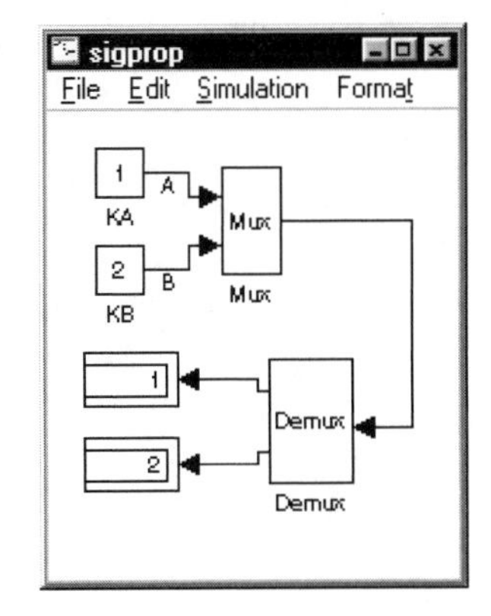

Label each signal line onto which you want the labels propagated with the single character <. Here we have configured all three signal lines to propagate the labels.

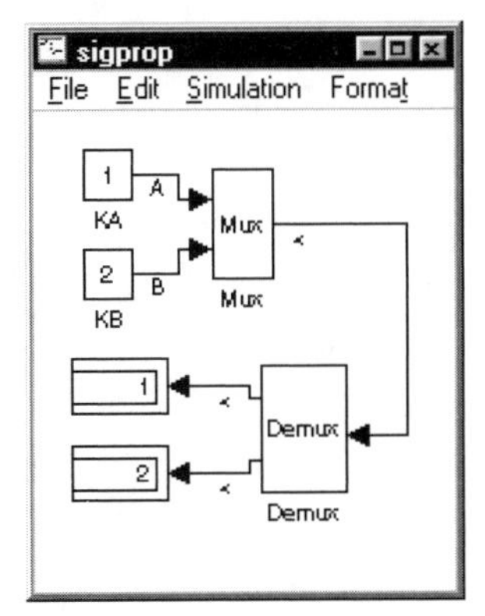

Choose **Edit:Update Diagram** from the menu bar of the model window. Here the signal line leaving the Mux block contains two signals, so the label is changed to `<A,B>`, showing both. The Demux block separates the vector signal into components, so the signal lines leaving it each contain only one signal, as shown by the labels. In both cases the propagated labels are enclosed in angle brackets to distinguish them from simple labels.

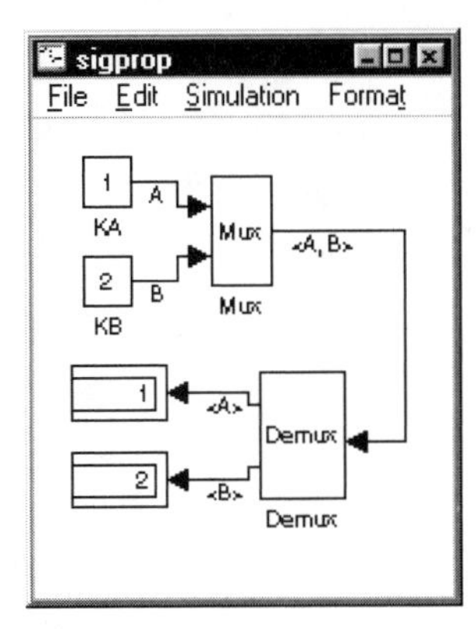

3.5 Annotations

You can add annotations to a model to make it easier to understand. You can also change the font used in an annotation to add emphasis.

3.5.1 Adding Annotations

Double click at the location where you want the center of an annotation. An editing cursor will appear.

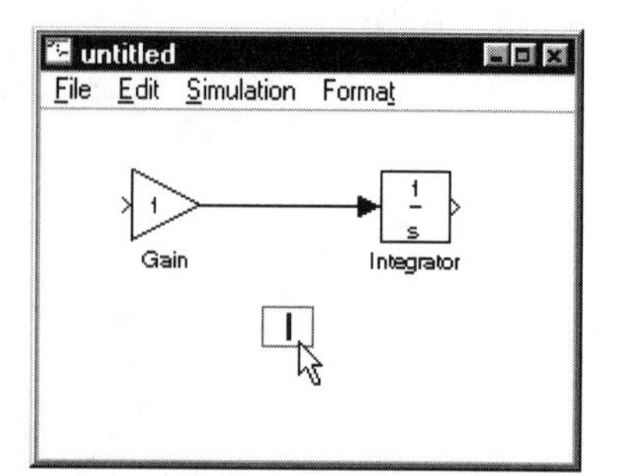

Enter the annotation. Press Return at the end of each line of a multiple line annotation. Click away from the annotation to complete the process.

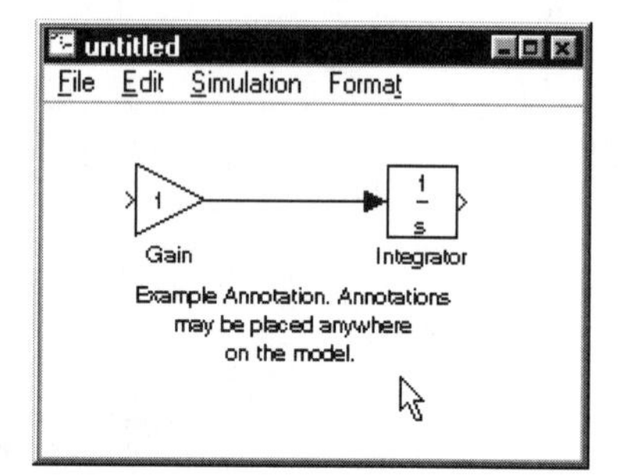

You can move and copy annotations using the same procedures used to move and copy blocks.

3.5.2 Changing Annotation Fonts

To change the font of an annotation, select the annotation. Choose **Format:Font** from the menu bar of the model window. A font selection dialog box will be displayed. Select the desired font, then press **OK**, and then click away from the annotation. All characters in a particular annotation will be the same font, but different annotations can be in different fonts.

3.6 Adding Sources

The inputs to a model are called sources, located in the Sources block library. A source block has no inputs and at least one output. Detailed documentation for each block in the Sources block library is available via the Block Browser. In this section we will mention several of the more commonly used blocks in the Sources block library. Then we will briefly discuss From Workspace and From File blocks, which allow you to create any signal you can describe mathematically in MATLAB and then use that signal as a SIMULINK input.

3.6.1 Common Sources

Many of the input signals used in modeling dynamical systems are available in the Sources block library. The Constant block produces a fixed constant signal, the magnitude of which is set in the block dialog box and displayed on the block icon. The Step block produces a step function. You can set the time of occurrence of the step, and the signal magnitude before and after the step. There is a Sine Wave block for which you can set the amplitude, phase, and frequency. The Signal Generator block can be set to produce sine, square, sawtooth, or random signals. More complex signals can be generated by combining the signals from multiple source blocks using a Sum block.

Example 3-1

A signal that is very useful in determining the behavior of dynamical systems is the *unit impulse*, also known as the delta function or Dirac delta function. (For a detailed discussion, see Meirovitch [1].) The unit impulse ($\delta(t-a)$) is defined to be a signal of zero duration, having the properties

$$\delta(t-a) = 0, \quad t \neq a$$

$$\int_{-\infty}^{\infty} \delta(t)dt = 1$$

Although the unit impulse is a theoretical signal that can't exist, it is a close approximation to real impulse signals, which are very common. Physical examples are collisions, such as a wheel hitting a curb or a bat hitting a ball, or near instantaneous velocity changes, such as firing a bullet from a rifle. Another use for the unit impulse is the assessment of a system's dynamics. The motion of a system forced by a unit impulse is due purely to the dynamics inherent in the system. Thus you can use the impulse response of a complex system to determine its natural frequencies and modes of vibration.

You can approximate a unit impulse function using two Step blocks and a Sum block, as shown in Figure 3-2. The idea is to produce, at the desired time (a), a very short-duration (d) pulse of a magnitude (M) such that $Md = 1$. The trick is in deciding on the proper value of d. It must be short relative to the fastest dynamics of the system. If it is too short, it can cause numerical problems, such as excessive roundoff error. If it is too long, it doesn't adequately simulate a

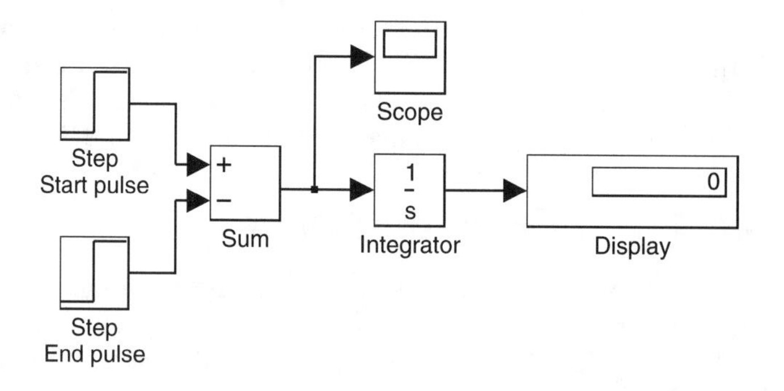

Figure 3-2 Generating a unit impulse function

true impulse. Usually, an acceptable compromise can be found with a little experimentation.

The model in Figure 3-2 is set to simulate a unit impulse at 0.5 seconds with a pulse of 0.01 second duration and magnitude of 100. The Step block labeled Step Start pulse is configured as follows: **Step time** is `0.5`, **Initial value** is `0`, **Final value** is `100`. The Step block labeled Step End pulse is configured as follows: **Step time** is `0.51`, **Initial value** is `0`, **Final value** is `100`. The simulation is configured to stop at 1 second. A plot of the output of the Sum block is shown in Figure 3-3. The Integrator block computes the time integral of the output of the Sum block, which is displayed using a Display block and which has the desired value (1).

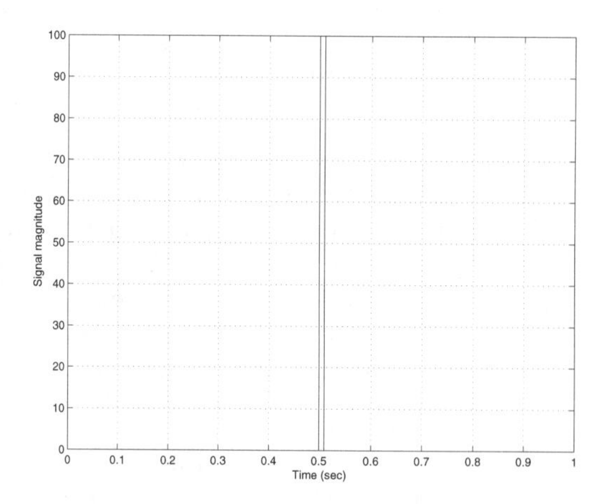

Figure 3-3 Unit impulse signal

3.6.2 From Workspace Block

The From Workspace block permits you to design a custom input signal. The block and its dialog box are illustrated in Figure 3-4. The block configuration parameter is a matrix table with the default value `[T,U]`. The input must be in the form of a MATLAB matrix, using variables currently defined in the MATLAB workspace. The first column of the matrix is the independent variable that corresponds to simulation time and must be monotonically increasing. The subsequent columns are values of the dependent variables corresponding to the independent variable in the first column. The block will produce as many outputs as there are dependent variables. The outputs are produced by linearly interpolating or extrapolating in the table.

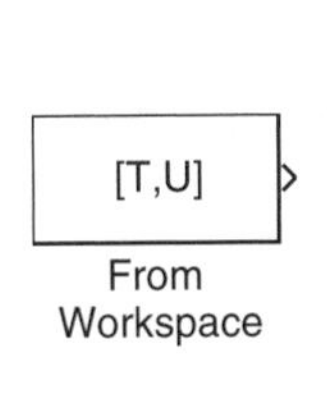

(a) From Workspace block

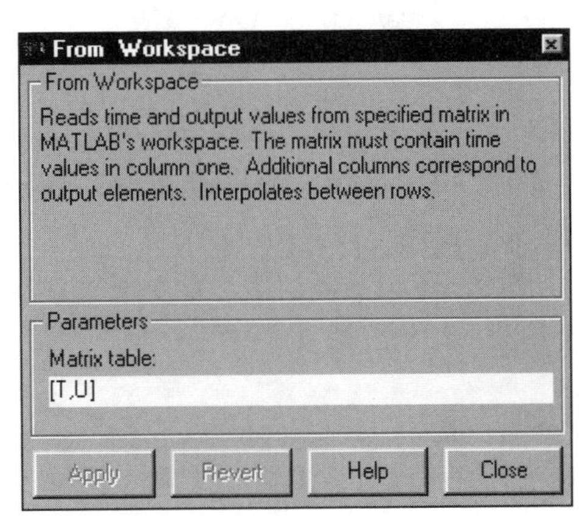

(b) From Workspace dialog box

Figure 3-4 From Workspace block and dialog box

Example 3-2

To illustrate the use of the From Workspace block, suppose we wish to generate a signal defined as

$$u(t) = t^2.$$

Figure 3-5 illustrates an M-file that produces a suitable input table.

Create the M-file using a text editor, and save it as a standard M-file using the `.m` extension. Note that the M-file must be saved with a name different from the SIMULINK model file. To use this table, you must first execute the M-file from within MATLAB, creating the table `A`. Configure the From Workspace

```
% Generate a signal for a From Workspace block
t = 0:0.1:100 ; % Independent variable
u = t.^2 ;      % Dependent variable
A = [t',u'] ;   % Form the table
```

Figure 3-5 M-file to create input table for From Workspace block

block by replacing `[T,U]` in the From Workspace dialog box with the name of the table (`A`).

Set **Matrix table** to `A`, the name of the table created in the MATLAB workspace. Click **Close**.

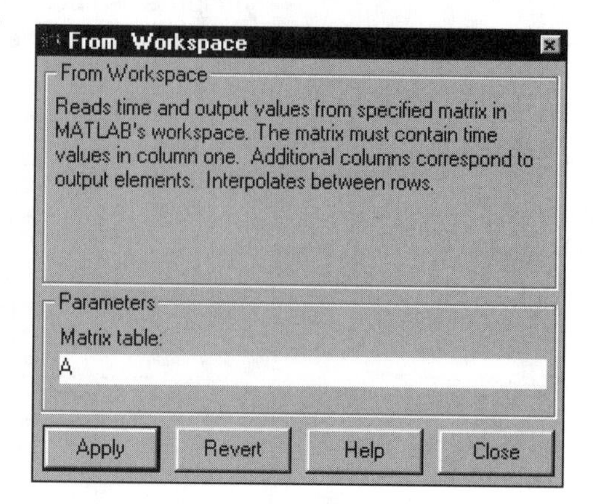

The block icon will display the name of the table.

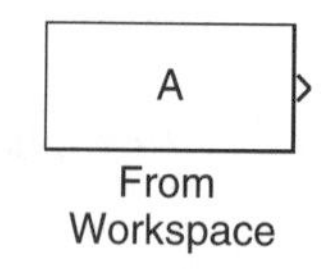

From File Input Block

The From File block is very similar to the From Workspace block. The primary difference is that the matrix is loaded from a MATLAB file (.mat) rather than coming directly from the MATLAB workspace. An additional difference is that the signals are stored in rows rather than in columns. You can produce a file (`examp.mat`) containing the matrix produced by the M-file in Figure 3-5 using the MATLAB commands

```
B = A'
save examp B
```

Configure the From File block by setting the full file name (for example, `examp.mat`) in the From File block dialog box field **File name**.

3.7 Adding Sinks

Sinks provide means to view or store model data. The Scope and XY Graph blocks produce plots of model data, and the Display block produces a digital display of the value of its input. The To Workspace block saves a signal to the MATLAB workspace, and the To File block saves a signal in MATLAB MAT-file format. The Stop block causes a simulation to stop when its input is nonzero. A detailed reference for each of these blocks is available in the Block Browser. We will discuss the Scope block and XY Graph block in some detail in this section.

3.7.1 Scope Block

The Scope block emulates an oscilloscope. The block shows a segment of the input signal, which may be either scalar or vector. Both the vertical range (y-axis) and horizontal range (time on the x-axis) can be set to any desired values. The vertical axis displays the actual value of the input signal. The horizontal axis scale always starts at zero and ends at the value specified as **Time range**. So, for example, if the horizontal range is 10 and the current time is 100, the input data for the period 90 to 100 is displayed, although the x-axis labels will still be 0 to 10. The Scope block is intended primarily for use during a simulation, and therefore Scope blocks do not provide a means to save the scope image for printing or inclusion in a document. However, the Scope block will send the signals it plots to the MATLAB workspace for further analysis or plotting using, for example, the MATLAB `plot` command.

You can place a Scope block in a model without connecting a signal line to the input of the Scope block and configure the block as a *floating Scope block*. A floating Scope block will use as its input any signal line that you click on during the execution of a simulation.

Figure 3-6 illustrates a Scope block. Note that there is a tool bar that contains six icons along the top of the Scope block window. These buttons allow you to zoom in on a portion of the display, autoscale the display, save a configuration for future use, and open a Scope properties dialog box. Table 3-1 describes the function of each button.

Zooming the Scope Display

Consider the model shown in Figure 3-7. The top Sine Wave block is configured to produce the signal $\sin(t)$ and the other Sine Wave block is configured to produce $0.4\sin(10t)$.

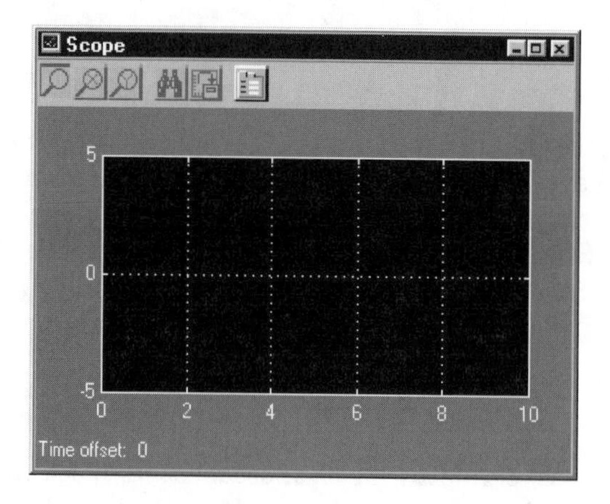

Figure 3-6 Scope block display before running simulation

Table 3-1 Scope block buttons

Button Icon	Function
	The **Zoom** button permits you to enlarge a region of the display.
	The **Zoom X** button allows you to zoom in on a portion of the display without changing the vertical scale.
	The **Zoom Y** button allows you to zoom in on a portion of the display without changing the horizontal scale.
	Autoscale changes the vertical scale such that the lower limit is the same as the minimum value in the currently displayed signal, and the upper limit is the same as the maximum value in the currently displayed signal. You can click on **Autoscale** during a simulation to rescale the display.
	Save axis makes the current scale the default for this Scope block. If you change the scale and then rerun the simulation without pressing **Save axis** first, the scale will revert to the current default when the simulation starts.
	Clicking on the **Open properties window** button opens the Scope properties dialog box. This dialog box allows you to set the default scales for the Scope block and to send the scope data to the MATLAB workspace.

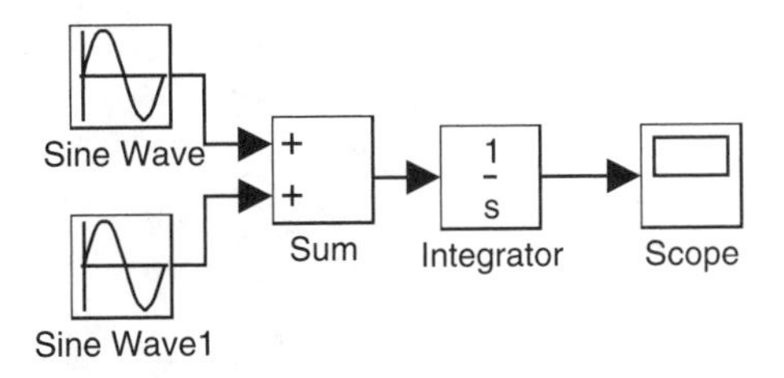

Figure 3-7 First-order system with sinusoidal input

Open the Scope block by double clicking on it. Run the simulation, resulting in the display shown here. At this point, you could rescale the display by clicking on the **Autoscale** button. Instead, zoom in on the peak between 2 and 4 seconds as follows.

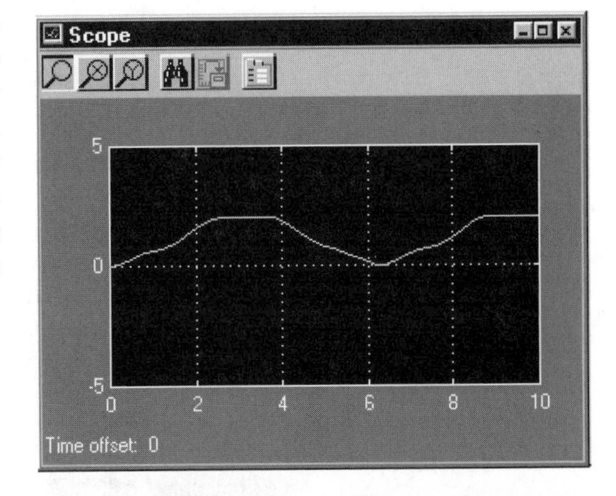

Click on the **Zoom** button. Then enclose the area you wish to zoom in on using a bounding box.

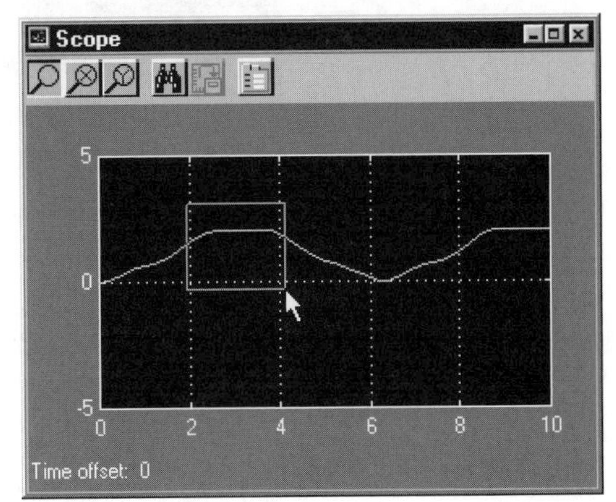

The Scope scales will change to include only the area you zoomed in on.

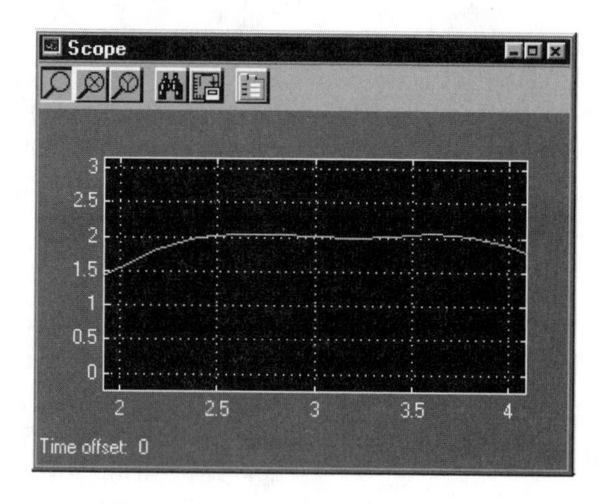

Scope Properties Dialog Box

The **Open properties window** button opens the Scope properties dialog box, which has two pages. The Axes page (Figure 3-8) has fields to enter the maximum (**Y max**) and minimum (**Y min**) values for the dependent variable. **Time range** may be set to a particular value or may be set to `auto`. If **Time range** is set to `auto`, the range will be the same as the simulation duration specified in the **Simulation:Parameters** dialog box.

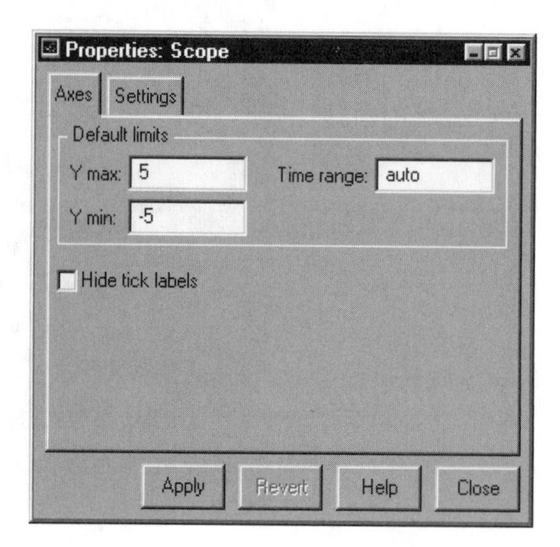

Figure 3-8 Scope properties axes page

The Settings page (Figure 3-9) has fields to control the number of points displayed and to save the data (not the display) to the MATLAB workspace. The

General section of the page consists of a drop-down list containing two choices: **Decimation** and **Sample time**. If **Decimation** is selected, the corresponding data field is set to a decimation factor that must be an integer. If **Decimation** is selected and set to 1 (the default), every point in the block input is plotted. If **Decimation** is set to 2, every other point is plotted, and so on. If **Sample time** is selected, the absolute spacing between plotted points must be entered in the data field.

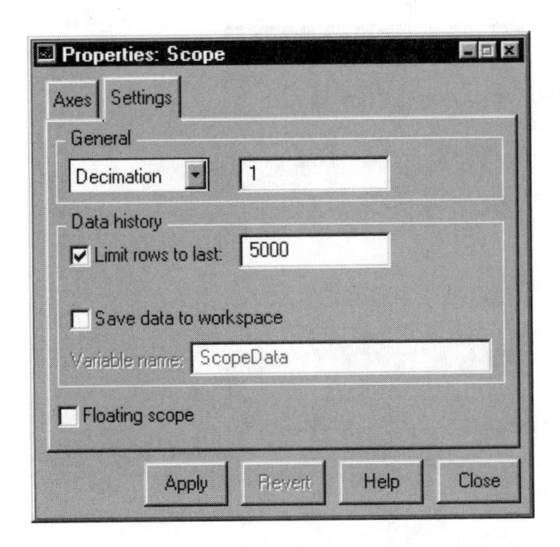

Figure 3-9 Scope properties settings page

The Scope block stores the input points in a buffer. Check **Limit rows to last** and enter a value to specify the size of the buffer (default is 5000). Autoscaling, zooming, and saving scope data to the workspace all work with this buffer. Thus if **Limit rows to last** is set to 1000 and the simulation produces a total of 2000 points, only the final 1000 points are available when the simulation stops.

As stated previously, there is no provision for printing the scope display or for embedding the scope display in a word processing document. However, you can send the scope data to the MATLAB workspace and use MATLAB's extensive plotting capabilities. To send the scope data to the MATLAB workspace, select **Save data to workspace** and enter the name of a MATLAB variable. When the simulation stops, the data displayed on the Scope will be stored in the MATLAB variable. There will be one column for the time values and one column for each signal input to the scope block. Thus if the signal entering the scope block is a vector signal with two components, the MATLAB variable will be a matrix with three columns and a number of rows equal to the number of time points displayed on the scope.

3.7.2 XY Graph

The XY Graph block produces a graph identical to a graph produced by the MATLAB command `plot`. The XY Graph accepts two scalar inputs. You must configure the horizontal and vertical ranges using the block dialog box. The XY Graph block dialog box is illustrated in Figure 3-10. The top input port is the x input, and the bottom input port is the y input.

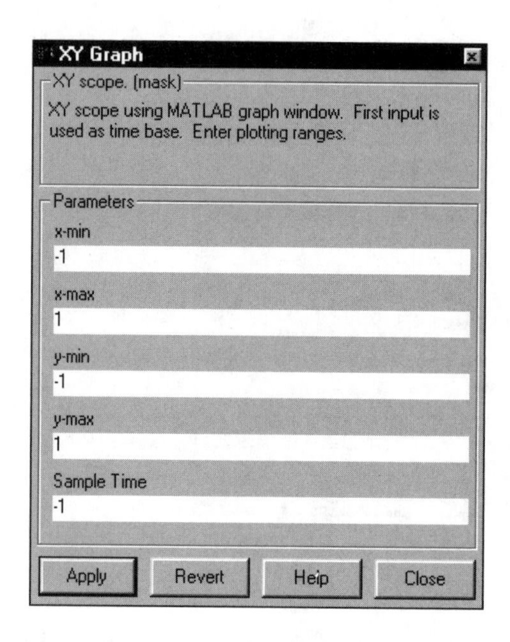

Figure 3-10 XY Graph block dialog box

3.8 Configuring the Simulation

A SIMULINK model is essentially a computer program that defines a set of differential and difference equations. When you choose **Simulation:Start** from the model window menu bar, SIMULINK solves that set of differential and difference equations numerically using one of its differential equation solvers. Before you run a simulation, you may set various simulation parameters, such as the starting and ending time, simulation step size, and various tolerances. You can choose among several high-quality integration algorithms. You can also configure SIMULINK to acquire certain data from the MATLAB workspace and to send simulation results to the MATLAB workspace.

Consider the model shown in Figure 3-11.

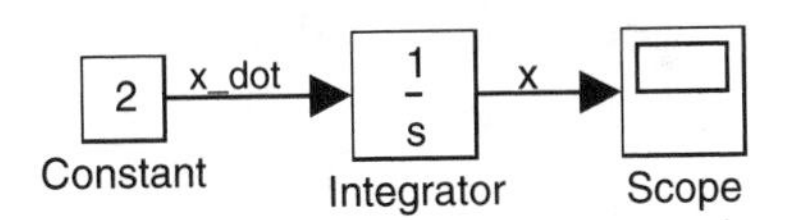

Figure 3-11 SIMULINK model of first-order system

This model represents the differential equation

$$\dot{x} = 2 \tag{3-1}$$

We set the Integrator block **Initial condition** field to 1. Next we choose **Simulation:Parameters** from the model window menu bar and set **Start time** to 0 and Stop time to 5. Then we choose **Simulation:Start** from the model window menu bar. SIMULINK will numerically evaluate the value of the integral to solve

$$x(\tau) = 1 + \int_0^{\tau} 2\,dt \tag{3-2}$$

and plot the value of $x(\tau)$ along the interval from 0 to 5. A numerical integration algorithm that solves this kind of problem (frequently called an *initial value problem*) is referred to as an *ordinary differential equation solver*, or just *solver* for the sake of brevity.

To set simulation parameters, choose **Simulation:Parameters** from the menu bar of the model window, which opens the Simulation parameters dialog box (Figure 3-12). The Simulation parameters dialog box contains three tabbed pages: Solver, Workspace I/O, and Diagnostics. The Solver page selects and configures the differential equation solver. The Workspace I/O page contains optional parameters that permit you to acquire simulation initialization data from the MATLAB workspace and to send certain simulation data to the MATLAB workspace. The Diagnostics page is used to select diagnostic modes, which are useful for troubleshooting certain simulation problems. We'll discuss each page in detail.

3.8.1 Solver Page

The Solver page consists of three sections. The first, Simulation time, contains fields to enter the start and stop times. **Start time** defaults to 0, and **Stop time** defaults to 10. The Solver options section contains fields to select the differential equation solver (numerical integration algorithm) and to set parameters that control the integration step size. The solvers are grouped in two categories: variable-step and fixed-step. Several different integration algorithms are available for each category. If a variable-step solver is chosen, there are fields to select the maximum integration step size, the initial integration step size, and absolute and relative tolerances. If a fixed-step solver is chosen,

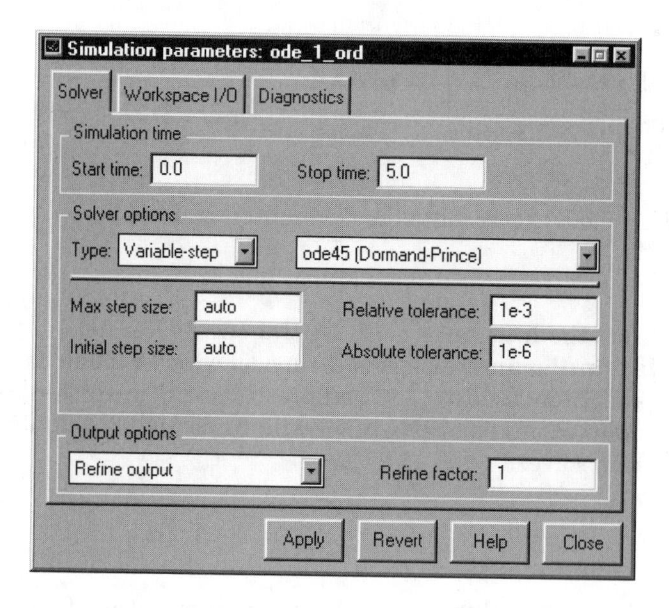

Figure 3-12 Simulation parameters dialog box

there is a single field in which to enter the step size. The Output options section controls the time spacing of points in the simulation output trajectory.

Solver Type

SIMULINK provides several ordinary differential equation solvers. The majority of these solvers are the result of recent numerical integration research and are among the fastest and most accurate methods available. Detailed descriptions of the algorithms are available in the paper by Shampine [2], available from The MathWorks.

It is generally best to use the variable-step solvers, as they continuously adjust the integration step size to maximize efficiency while maintaining a specified accuracy. SIMULINK's variable-step solvers can completely decouple the integration step size and the interval between output points, so it is not necessary to limit the step size to get a smooth plot or to produce an output trajectory with a predetermined fixed-step size. The available solvers are listed in Table 3-2. We will discuss them in more detail in Chapter 8.

Output options

The Output options section of the Solver page works in conjunction with the variable-step solvers to control the spacing between points in the output trajectory. Output options do not apply to the fixed-step solvers. The **Output**

Table 3-2 SIMULINK Solvers

Solver Type	Characteristics
ODE45	Excellent general-purpose single-step solver. Based on the Dormand-Prince fourth/fifth-order Runge-Kutta pair. ODE45 is the default solver and is usually a good first choice.
ODE23	Uses the Bogacki-Shampine second/third-order Runge-Kutta pair. Sometimes works better than ODE45 in the presence of mild stiffness. Generally requires a smaller step size than ODE45 to get the same accuracy.
ODE113	Variable-order Adams-Bashforth-Moulton solver. Since ODE113 uses the solutions at several previous time points to compute the solution at the current time point, it may produce the same accuracy as ODE45 or ODE23 with fewer derivative evaluations and thus perform much faster. Not suitable for systems with discontinuities.
ODE15S	Variable-order multistep solver for stiff systems. Based on recent research using numerical difference formulas. If a simulation runs extremely slowly using ODE45, try ODE15S.
ODE23S	Fixed-order single-step solver for stiff systems. Because ODE23S is a single-step method, it is sometimes faster than ODE15S. If a system appears to be stiff, it is a good idea to try both stiff solvers to determine which performs the best.
Fixed- and Variable-Step Discrete	Special solvers for systems that contain no continuous states.
ODE5	Fixed-step version of ODE45.
ODE4	Classic fourth-order Runge-Kutta formulas using a fixed step size.
ODE3	Fixed-step version of ODE23.
ODE2	Fixed-step second-order Runge-Kutta method, also known as Heun's method.
ODE1	Euler's method using a fixed step size.

options field contains a list box with three choices: **Refine output**, **Produce additional output**, and **Produce specified output only**.

Choose **Refine output** to force the solver to add intermediate points between the solution points for successive integration steps. SIMULINK computes the intermediate points using interpolation, which is much faster than using reduced integration step size. **Refine output** is a good choice if the output trajectory needs to appear smoother but there is no need for a fixed spacing between points. If **Refine output** is chosen, there will be an additional input field labeled **Refine factor**. **Refine factor** must be an integer. SIMULINK divides each integration step into **Refine factor** output steps; so, for example, if **Refine factor** is set to 2, the midpoint of each integration step will be added to the output trajectory.

Produce additional output permits you to force SIMULINK to include certain time points in the output trajectory, in addition to the solution points at the end of each integration step. If **Produce additional output** is selected, there will be an additional field labeled **Output times.** This field must contain a vector listing the additional times for which output is requested. For example, if it is necessary to include the output at 10-second intervals and the value of **Start time** is 0 and **Stop time** is 100, **Output times** should contain `[0:10:100]`.

Choose **Produce specified output only** if it is necessary to produce an output trajectory containing only specified time points. For example, you may wish to compare several trajectories to evaluate the effect of changing a parameter. If **Produce specified output only** is selected, there will be an additional field labeled **Output times**, which must contain a vector of the desired output times.

3.8.2 Workspace I/O Page

The Workspace I/O page (Figure 3-13), permits you to acquire simulation input from the MATLAB workspace and to send output directly to the MATLAB workspace. The page consists of four sections: Load from workspace, Save to workspace, States, and Save options. We'll discuss each section in detail.

Load from workspace

Selecting the **Load from workspace, Input** check box causes SIMULINK to take the input time points and values of the input variables from the MATLAB workspace. **Load from workspace** works in conjunction with the Inport block found on the Connections block library. Inport blocks may be configured to accept scalar or vector data. Set the name of the time and input matrixes in the **Load from workspace, Input** field. The first matrix (default name `t`) is a column vector of time values, and the second matrix (default name `u`) consists

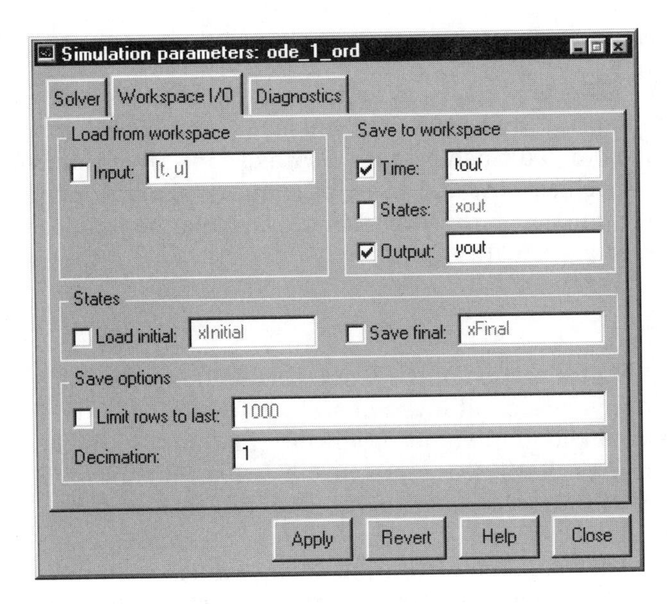

Figure 3-13 Workspace I/O page

of one column for each input variable, with a row corresponding to each row in the time matrix. If there is more than one Inport block, the columns of the input matrix are ordered corresponding to the number assigned to the Inport blocks. So the first column corresponds to the lowest numbered Inport block, and the last column corresponds to the highest numbered Inport block. For a vector Inport block, there must be one column in the u matrix for each element of the input vector.

SIMULINK Internal State Vectors

Before discussing the Save to workspace and States sections of the Workspace I/O page, we should briefly discuss SIMULINK's internal state variables. A SIMULINK model can be thought of as a set of simultaneous first-order, possibly nonlinear, differential and difference equations. In addition to the state variables associated with each integrator block, there are implicitly specified state variables associated with transfer function blocks, state-space blocks, certain nonlinear blocks, certain discrete blocks, and many of the blocks in the Extras block library. It is frequently useful to have access to a model's state variables, and SIMULINK provides mechanisms to facilitate this. Use of the Workspace I/O page is probably the easiest method to access a model's state variables. Accessing a model's state variables and, in particular, identifying all of a model's state variables is discussed further in Chapter 7.

Save to workspace

The Save to workspace section contains three fields, each activated with a check box. **Time** sends the independent variable to the specified workspace matrix (default name `tout`). **States** sends all of the model's state variables to the specified MATLAB workspace matrix (default name `xout`). **Output** works in conjunction with Outport blocks in a manner analogous to Inport blocks, discussed previously.

States

The States section of the Workspace I/O page can force SIMULINK to load the initial values of all internal state variables from the MATLAB workspace and to send the final values of all internal state variables to the MATLAB workspace. All SIMULINK state variables have default initial values, in most cases 0. States associated with Integrator blocks may be initialized to any value using the block's dialog box. Specifying initial states on the Workspace I/O page overrides any default initialization values, including initial values set in an Integrator block's dialog box. **States, Load initial** sets the initial value of a model's state vector to the values in the specified input vector (default name `xInitial`), defined in the MATLAB workspace. The initialization vector must be of the same size as the model's state vector. **States, Save final** saves the final value of the model's state vector in the specified matrix (default name `xFinal`) in the MATLAB workspace. The output of **States, Save final** is a suitable initial vector for **States, Load initial** and may be used to restart a model at the final point of a previous simulation.

Save options

The Save options section of the Workspace I/O page contains two fields and works in conjunction with the Save to workspace section. The first field, **Limit rows to last**, sends at most the specified number of points to the MATLAB workspace. So, for example, if **Limit rows to last** is checked and set to the default value of `1000`, at most the final 1000 points will be sent to the workspace. **Decimation** sets the interval between points sent to the MATLAB workspace. If **Decimation** is set to `1`, every point will be sent the workspace. If **Decimation** is set to `2`, every other point will be sent to the workspace, and so on. **Decimation** must be set to an integer value.

3.8.3 Diagnostics Page

The Diagnostics page (Figure 3-14), allows you to select the action taken for five exceptional conditions and also includes options to control automatic block output consistency checking and to disable zero crossing detection. There are three choices for the response to each of the five exception conditions. The first choice, **None**, instructs SIMULINK to ignore the corresponding exception. The second choice, **Warning** (the default choice), causes SIMULINK to issue a warning message each time the corresponding exception occurs. The final choice, **Error**, causes SIMULINK to abort the simulation and issue an error message whenever the corresponding exception occurs. The exceptions are detected when a model is executed. We'll briefly discuss each exceptional condition.

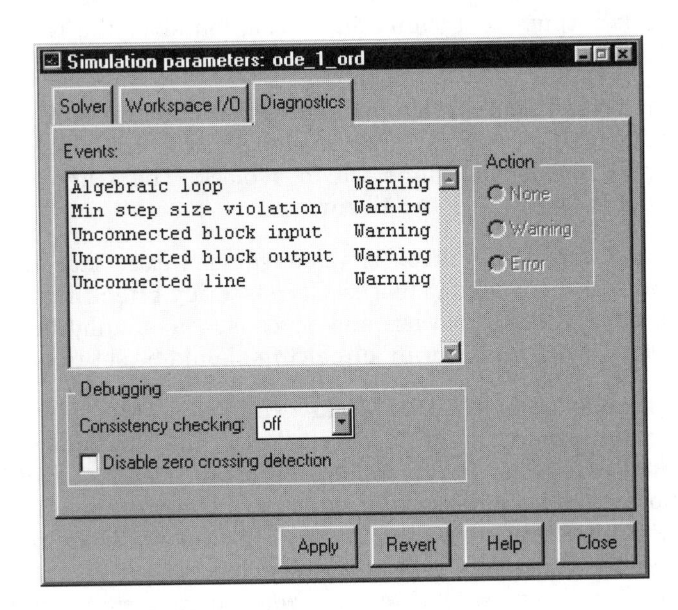

Figure 3-14 Diagnostics page

An **Algebraic loop** is an exception in which a block's input at a given instant of time is dependent on the same block's output at the same instant of time. Algebraic loops are troublesome because they can significantly reduce the speed of a simulation and, in some cases, can cause the simulation to fail to execute. A detailed discussion of algebraic loops is presented in Chapter 8. It is usually best to set the algebraic loop response to **Warning**. If an algebraic loop is discovered, and if performance is acceptable, change the response to **None**.

A **Min step size violation** occurs when the solver attempts to use an integration step size smaller than the minimum. It is not possible to change the minimum step size for any of the variable-step solvers. If this exception occurs, you can change to a higher-order solver, which in general will use a larger integration step size. Your other choice is to increase the absolute and relative tolerances on the Solver page. **Min step size violation** should always be set to **Warning** or **Error** because it indicates that the simulation is not producing the expected accuracy.

An **Unconnected block input** exception occurs when a block has an input that is not used. This is generally the result of an error in building the model. This exception should be set to **Error** or **Warning**. If the omission of a block's input is intentional, it is a good practice to connect a Ground block to the input.

An **Unconnected block output** exception occurs when a block has an output that is not connected to the input of another block. This exception is frequently harmless, but it is easy to solve. To prevent this exception, connect unused block outputs to Terminator blocks, found in the Connections block library. This exception should be set to **Warning** or **Error**.

The final exception, **Unconnected line**, is almost always due to an error in constructing the SIMULINK model. This exception occurs when one end of a signal line is not connected to a block. The **Unconnected line** exception should usually be set to **Error**.

Consistency checking is a debugging feature that detects certain programming errors in custom blocks. **Consistency checking** is not needed with standard SIMULINK blocks, and it causes a simulation to run much slower. Ordinarily, **Consistency checking** should be set to **off**.

A number of SIMULINK blocks exhibit discontinuous behavior. For example, the output of the Sign block, located in the Nonlinear block library, is 1 if its input is positive, 0 if its input is zero, and –1 if its input is negative. Thus, the block exhibits a discontinuity at zero. If a variable-step-size solver is in use, SIMULINK will adjust the integration step size when the input to a Sign block is approaching zero so that the switch occurs at the right time. This process is called *zero crossing detection*. You can determine if a block invokes zero crossing detection by referring to the Characteristics table for the block in the Block Browser.

Zero crossing detection improves the accuracy of a simulation, but it can cause a simulation to run slowly. Occasionally, a system will fluctuate rapidly about a discontinuity, a phenomenon called *chatter*. When this happens, the progress of the simulation can effectively stop as the integration step size is reduced to a very small value. If your model runs very slowly and includes one or more blocks with intrinsic zero crossing detection, selecting **Disable zero crossing detection** on the Diagnostics page can significantly increase the speed of a simulation. However, this can also adversely affect the accuracy of a simulation, so it is best used only as a tool to verify that chatter is occurring. If selecting **Disable zero crossing detection** dramatically improves the speed of a simulation, you should locate the cause of chatter and correct the problem.

3.9 Running a Simulation

You can control the execution of the model using the **Simulation** pull-down menu on the model window menu bar. To start the simulation, select **Simulation:Start**. You can stop the simulation at any time using **Simulation:Stop**. Choose **Simulation:Pause** to halt execution temporarily. Then choose **Simulation:Continue** to resume or **Simulation:Stop** to halt execution of the model permanently. It is also possible to run a simulation from the MATLAB command line; this method will be discussed in Chapter 7.

While the simulation is executing, you can change many parameters. For example, you can change the gain of a gain block, choose a different solver, or change integration parameters such as minimum step size. You can also select a signal line that will become the input to a floating Scope block. This permits you to check various signals as the simulation progresses.

3.10 Printing a Model

There are a variety of options for printing SIMULINK models. The simplest is to send the model directly to a printer. However, it is also possible to embed the model in a word processing document or other file, either by copying it to the clipboard or saving it as an Encapsulated PostScript file.

3.10.1 Printing to the Printer Using Menus

The fastest way to get a printed copy of a model is to send it directly to the printer using the menus. To do this, choose **File:Printer Setup** from the model window menu bar and configure the printer as desired. Next choose **File:Print**. When you send a model directly to the printer, SIMULINK resizes it as necessary to fit on the page. You have no control over the size of the model.

3.10.2 Embedding the Model in a Document

Modern documentation applications, such as word processors, presentation programs, desktop publishing programs, drawing programs, and even spreadsheets, allow you to insert SIMULINK model images in documents.

On Windows computers, SIMULINK model images may be embedded as bitmaps, Windows metafiles, or Encapsulated PostScript (EPS) files. To embed an image as a bitmap or metafile, first select **Edit:Copy Model** from the model window menu bar to copy the image. Next make the target document active, and use the procedure for the documenting application to embed the image. Generally, it is preferable to use the Metafile format, because this format does not lose resolution when resized. However, not all programs handle Metafiles correctly, and the bitmap option provides a reliable alternative. To embed an image in EPS form, first save the image as an EPS file using the `print` command, which will be discussed shortly. Next make the target document active, and use the procedure for the documenting application to import the file (refer to the application's documentation if necessary). Note that if you use EPS, you will probably have to use a PostScript printer in order for the image to appear in hardcopy output.

On Macintosh computers, SIMULINK model images may be embedded via the Clipboard or as EPS files. To embed an image via the Clipboard, first choose **Edit:Copy** from the model window menu bar to copy the image. Next make the target document active, and choose **Edit:Paste** to embed the image. To embed an image in EPS form, first save the image as an EPS file by choosing **File:Save As...** and filling in the dialog box to save the image to file. Next make the target document active, and use the procedure for the documenting application to import the file (refer to the application's documentation if necessary).

3.10.3 Using the MATLAB `print` Command

The MATLAB `print` command permits you to send a model image to a printer, to the clipboard, or to a file in a variety of formats. The syntax of the `print` command is

```
print -smodel -ddevice filename
```

`model` is a MATLAB string containing the name of a currently open SIMULINK model. The model name is displayed in the title area of the model window. Note that filenames on your computer may be case sensitive. If the model name contains spaces, enclose the name in single quotes:

```
print -s'Spring Mass System' -ddevice filename
```

If the model name contains a carriage return—that is, if it's shown in the window title in two lines—represent the carriage return as its ASCII code (13) and enclose each line in single quotes ('), and the whole name in brackets (`[]`):

```
print -s['Damped' 13 'Spring Mass System'] -ddevice filename
```

`device` is a MATLAB string that specifies the type of output device. Devices include printers, files, and the Windows clipboard. Table 3-3 lists the available device types.

`filename` is a MATLAB string containing a valid file name and is an optional argument. If *`filename`* is specified, the output will be directed to the specified file rather than the printer. If *`device`* is an Encapsulated PostScript format and *`filename`* is not specified, *`filename`* will default to `Untitled.eps`.

Table 3-3 Device Codes for the `print` command

device	**Device Description**
ps	PostScript
psc	Color PostScript
ps2	Level 2 PostScript
psc2	Level 2 Color PostScript
eps	Encapsulated PostScript (must go to a file)
epsc	Color Encapsulated PostScript (must go to a file)
eps2	Encapsulated Level 2 PostScript (must go to a file)
epsc2	Color Encapsulated Level 2 PostScript (must go to a file)
win	Current printer
winc	Current printer, color
meta	Clipboard in Metafile format
bitmap	Clipboard in bitmap format
setup	Same as selecting **File:Printer Setup** from the model window's menu bar

Example 3-3

In this example we will, from the MATLAB prompt, print a model named `xydemo.mdl` to the current default printer and then save the model image as a color Encapsulated PostScript (EPS) file.

First open model `xydemo.mdl`. Next, at the MATLAB prompt, enter the following command:

```
print -s'xydemo'
```

The model will be printed. Next enter the command

```
print -s'xydemo' -depsc xydemo.eps
```

The model is now saved to file. It can be imported into an application that accepts eps files.

3.11 Model-Building Summary

Table 3-4 summarizes the model-building procedures discussed in Chapter 2 and Chapter 3. The following terms are used:

Drag	Click and hold the left mouse button; drag object to new location
Shift-click	Holding down the **Shift** key, click with the left mouse button.
Shift-drag	Holding down the **Shift** key, drag with the left mouse button.
Control-drag	Holding down the **Ctrl** key, drag with the left mouse button (Windows only).
Right-drag	Drag using the right mouse button (Windows only).

Table 3-4 Summary of Model Building Operations

Operation	Windows	Macintosh
Select object (block or signal line)	Click on the object with the left mouse button.	Click on the object with the mouse button.
Select another object	Shift-click the additional object.	Shift-click the additional object.
Select with bounding box	Click with the left mouse button at the location of one corner of the bounding box. Continuing to depress the mouse button, drag the bounding box to enclose the desired area.	Click with the mouse button at the location of one corner of the bounding box. Continuing to depress the mouse button, drag the bounding box to enclose the desired area.
Copy block from block library or another model	Select block, then drag it to the model window.	Select block, then drag it to the model window.
Flip block	Select block, then **Format:Flip Block**. Shortcut: **Ctrl-f**.	Select block, then **Format:Flip Block**. Shortcut: **⌘-f**
Rotate block	Select block, then **Format:Rotate Block**. Shortcut: **Ctrl-r**.	Select block, then **Format:Rotate Block**. Shortcut: **⌘-r**
Resize block	Select block, then drag handle.	Select block, then drag handle.
Add drop shadow	Select block, then **Format:Show Drop Shadow**.	Select block, then **Format:Show Drop Shadow**.
Edit block name	Click on name.	Click on name.

Table 3-4 Summary of Model Building Operations (Continued)

Operation	**Windows**	**Macintosh**
Hide block name	Select name, then **Format:Hide Name**.	Select name, then **Format:Hide Name**.
Flip block name	Select name, then **Format:Flip Name**. Shortcut: Drag name to new location.	Select name, then **Format:Flip Name**. Shortcut: Drag name to new location.
Delete object	Select object, then **Edit:Clear**. Shortcut: **Delete** key.	Select object, then **Edit:Clear**. Shortcut: **Delete** key.
Copy object to clipboard	Select object, then **Edit:Copy**. Shortcut: **Ctrl-c**.	Select object, then **Edit:Copy**. Shortcut: **⌘-c**.
Cut object to clipboard	Select object, then **Edit:Cut**. Shortcut: **Ctrl-x**.	Select object, then **Edit:Cut**. Shortcut: **⌘-x**.
Paste from clipboard	**Edit:Paste**. Shortcut: **Ctrl-v**.	**Edit:Paste**. Shortcut: **⌘-v**.
Draw signal line	Drag from output port to input port	Drag from output port to input port
Draw signal line in segments	Drag from output port to first bend. Release mouse button. Drag from first bend to second bend, and so on.	Drag from output port to first bend. Release mouse button. Drag from first bend to second bend, and so on.
Draw signal line at arbitrary angle	Shift-drag from output port to input port.	Shift-drag from output port to input port.
Branch from signal line	Control-drag from point of branch. Shortcut: Right-drag, starting from the point of branch.	Option-drag from the point of branch.
Split signal line	Select line. Shift-drag the new vertex.	Select line. Shift-drag the new vertex.
Move line segment	Drag segment.	Drag segment.
Move line segment vertex	Drag vertex.	Drag vertex.
Label signal line	Double click on line, and type text.	Double click on line, and type text.
Move signal line label	Drag the label to the desired location.	Drag the label to the desired location.

Table 3-4 Summary of Model Building Operations (Continued)

Operation	Windows	Macintosh
Copy signal line label	Control-drag the copy of the label to desired location. Shortcut: Right-drag to desired location.	Option-drag the copy of the label to desired location.
Delete one occurrence of signal line label that has multiple occurrences	Shift-click label, then press **Delete** key.	Shift-click label, then press **Delete** key.
Delete all occurrences of signal line label	Select label, then delete all characters in the label.	Select label, then delete all characters in the label.
Propagate signal label	Label signal line onto which you want label propagated with single character <. Then choose **Edit:Update Diagram**.	Label signal line onto which you want label propagated with single character <. Then choose **Edit:Update Diagram**.
Add annotation to model	Double click at location of annotation, and type text.	Double click at location of annotation, and type text.

3.12 Summary

In this chapter we have described procedures for editing and annotating SIMULINK models. We have also discussed configuring and printing a SIMULINK model. In the next two chapters we will discuss the use of these procedures to model continuous, discrete, and hybrid systems.

3.13 References

1 Meirovitch, Leonard, *Introduction to Dynamics and Control*. New York: John Wiley & Sons, 1985, pp 16–17.

2 Shampine, Lawrence F., and Reichelt, Mark W., "The MATLAB ODE Suite," The MathWorks, Inc., Natick, MA, 1996. This technical paper is available directly from The MathWorks. It provides a detailed discussion of the MATLAB differential equation solvers that are available from within SIMULINK.

4

Continuous Systems

In this chapter, we will discuss the use of SIMULINK to model continuous systems. We'll start with scalar linear systems. Then we'll model vector linear systems. Finally, we will use blocks from the Nonlinear block library to model nonlinear continuous systems.

4.1 Introduction

After studying Chapters 2 and 3 and experimenting a little, you should be comfortable with the mechanics of building and running SIMULINK models. In this chapter we will explore using SIMULINK to model continuous systems.

A continuous system is a dynamical system that can be described using differential equations. Thus most physical systems and processes are continuous. The simplest systems are scalar and can be assumed to be linear and time invariant. We'll start this chapter by modeling these simple systems using blocks from the Linear block library. Next we'll show how to model more complex systems using vector signals. We'll also discuss the use of blocks from the Nonlinear block library to model nonlinear continuous systems.

4.2 Scalar Linear Systems

Scalar linear systems can be modeled using blocks in the Linear block library. These blocks are easy to use, but the Integrator block has several important capabilities that merit elaboration. In this section we will start with a detailed description of the Integrator block. Next we will illustrate modeling scalar continuous systems with two examples.

4.2.1 Integrator Block

You can configure the Integrator block as a simple integrator or as a reset integrator. A reset integrator resets its output to the initial condition value when the reset signal triggers. You can also configure an Integrator block such that its output stays within preset limits. Additionally, you can set the integrator's initial output value in the Integrator block's dialog box, or configure the integrator to receive its initial output value through an additional input port.

Double clicking on an Integrator block opens the Integrator block dialog box, shown in Figure 4-1. To use a standard integrator, the only required input is **Initial condition**, which defaults to 0.

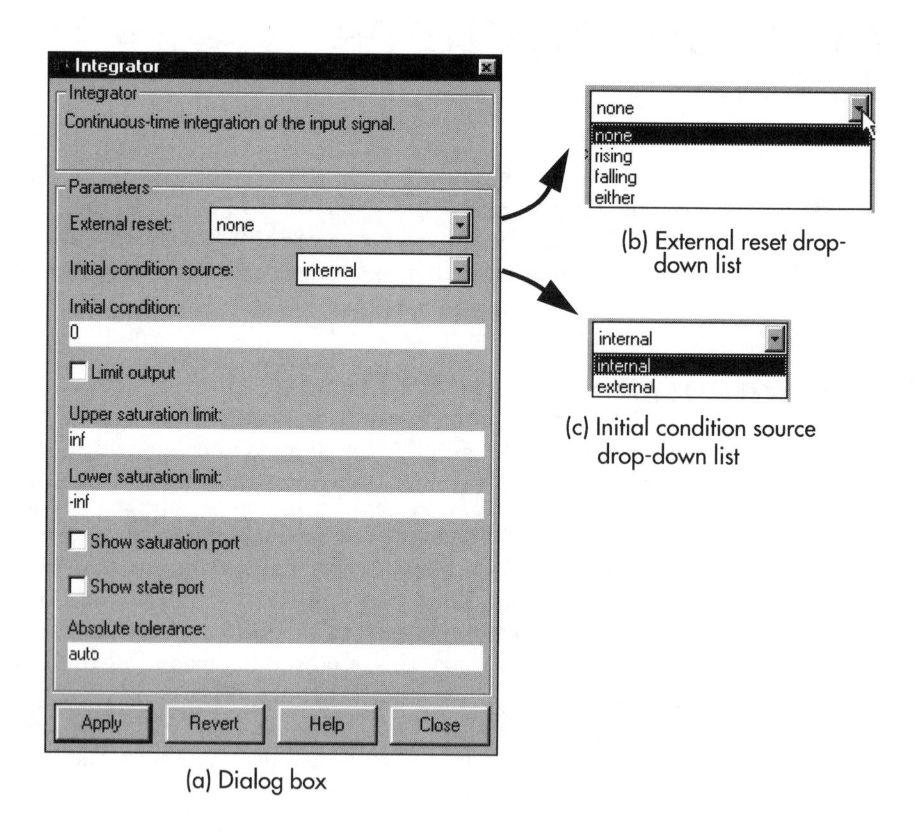

(a) Dialog box

(b) External reset drop-down list

(c) Initial condition source drop-down list

Figure 4-1 Integrator block dialog box

External reset is a drop-down list containing four choices: **none**, **rising**, **falling**, and **either**. Selecting **none** (the default) disables the external reset feature. If one of the other choices is selected, the Integrator block becomes a reset integrator block. If the reset choice is **rising**, the integrator output is reset to its initial condition value when the reset signal crosses (or departs rising if it starts at zero) zero from below. If set to **falling**, the output is reset to the initial condition value when the reset signal crosses zero descending. If set to **either**, the output is reset to the initial condition value when the reset signal crosses zero from above or below. Figure 4-2 illustrates a simple model with the integrator **External reset** set to **falling**. Figure 4-3 shows the output and reset trajectories. When the reset signal crosses zero, the integrator output is reset to its initial value and the simulation continues.

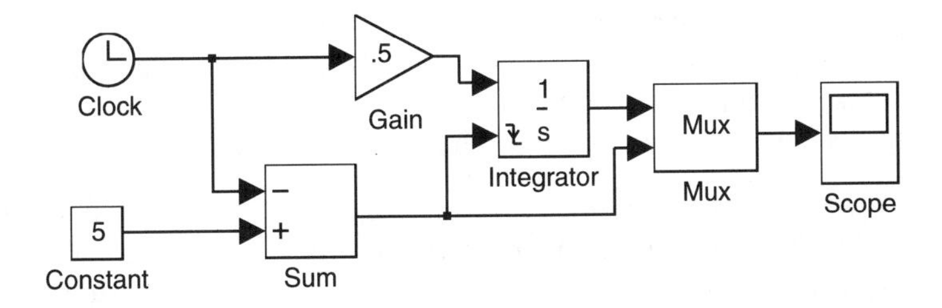

Figure 4-2 Model with reset integrator

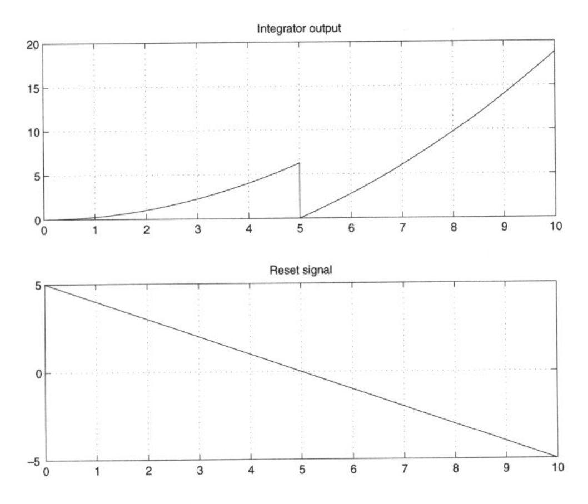

Figure 4-3 Reset integrator output and reset signals

Setting **Initial condition source** to **external** adds an additional input port to the integrator. The value of this input will be used as the initial output of the integrator when the integration starts and when the integrator is reset if an external reset is present.

Selecting **Limit output** causes the block to function as a limited integrator. The value of the output will be no greater than **Upper saturation limit**, and no lower than **Lower saturation limit**. The default upper saturation limit is `inf`, which represents infinity, and the default lower saturation limit is `-inf`.

Selecting **Show saturation port** adds an additional output port that indicates the saturation status. The signal from this port will be –1 if the lower saturation limit is active, 1 if the upper saturation limit is active, and 0 if the output is between the saturation limits.

Selecting **Show state port** adds an additional output port to the block. This port will output the integrator state, which is the same as the integrator output. There are two situations in which the state port is needed. If the output of an Integrator block is fed back into the same block's reset or initial condition port, the state port signal must be used instead of the block output. You should also use a state port if you want to pass the output of an Integrator block in a conditionally executed subsystem (see Chapter 6 for details on conditionally executed subsystems) to another conditionally executed subsystem.

Absolute tolerance allows you to override the **Absolute tolerance** setting in the **Simulation:Parameters** dialog box for the output of a particular Integrator block. See Chapter 8 for a detailed explanation of **Absolute tolerance**. In a situation in which the absolute value of the output of an Integrator block is different by several orders of magnitude from other signals in a model, setting **Absolute tolerance** for that Integrator block appropriately can improve the accuracy of the simulation.

Figure 4-4 shows an Integrator block with all possible ports active.

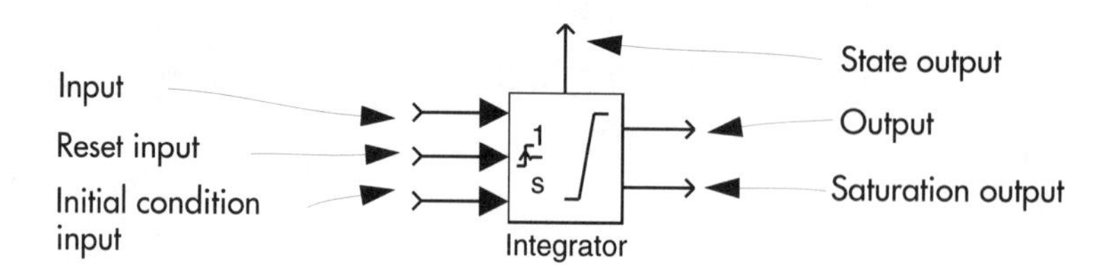

Figure 4-4 Integrator block with all options selected

Example 4-1

Consider the damped second-order system illustrated in Figure 4-5.

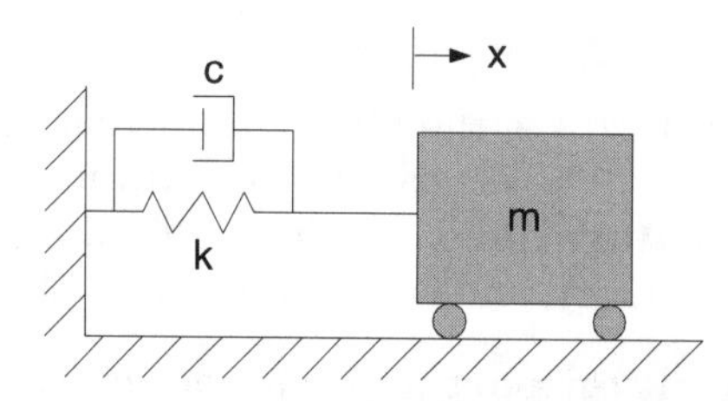

Figure 4-5 Damped second-order system

Assume that the damping coefficient $c = 1.0$, the spring constant $k = 2$ lb/ft, and the cart mass $m = 5$ slugs. There is no input to this system. We will model the motion of the cart, assuming it is initially deflected 1 ft from the equilibrium position.

To model the system, it is necessary to write the equation of motion. Using the Newtonian approach, we note that there are two forces acting on the cart: the spring force and the damping force. The spring force is kx, and the damping force is $c\dot{x}$. The force due to the acceleration of the cart is $m\ddot{x}$. Since there are no externally applied forces, the sum of these three forces must be zero. Thus we can write the equation of motion

$$m\ddot{x} + c\dot{x} + kx = 0$$

Since this is a second-order system, we will need two integrators to model its behavior. Open a new model window and then start building the model by dragging two integrators from the Linear block library. Label them as shown and connect the output of the `Velocity` integrator to the input of the `Displacement` integrator.

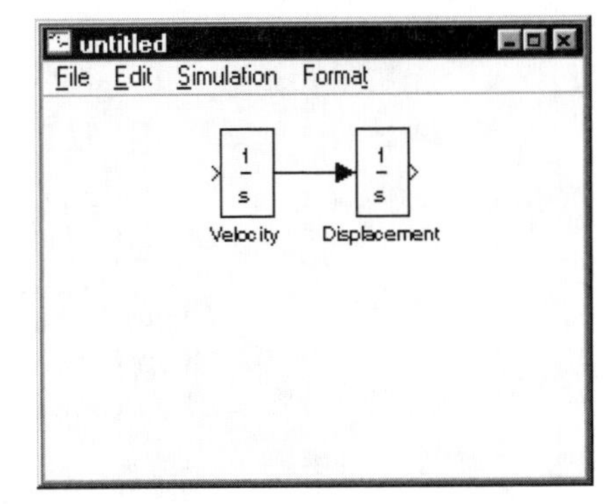

The output of the `Velocity` integrator is $\dot{x}$, so its input must be $\ddot{x}$. Rewrite the equation of motion to compute $\ddot{x}$ as a function of x and $\dot{x}$:

$$\ddot{x} = -\frac{c}{m}\dot{x} - \frac{k}{m}x$$

Substituting the values of the parameters,

$$\ddot{x} = -0.2\dot{x} - 0.4x$$

with $x(0) = 1$, $\dot{x}(0) = 0$.

Add a Sum block to compute $\ddot{x}$. The Sum block is configured with two minus signs (– –).

Add Gain blocks to compute the ratio of damping force to mass and the ratio of the spring force to mass. Draw signal lines as shown and label the Gain blocks to show their function more clearly.

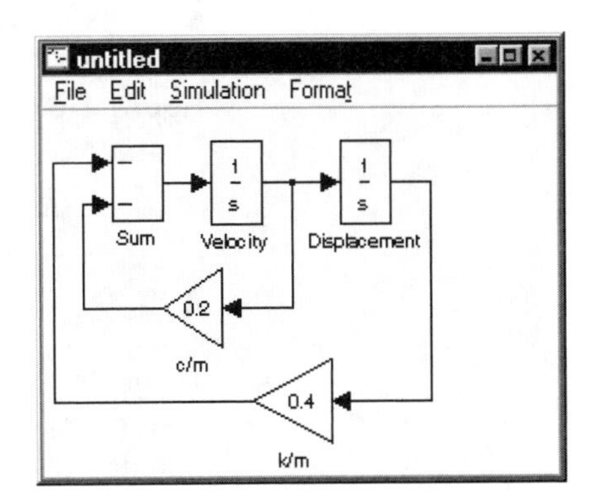

Set the **Initial condition** field in the `Velocity` integrator's dialog box to `0` and the **Initial condition** field in the `Displacement` integrator's dialog box to `1`.

Add a Scope block to display the block position.

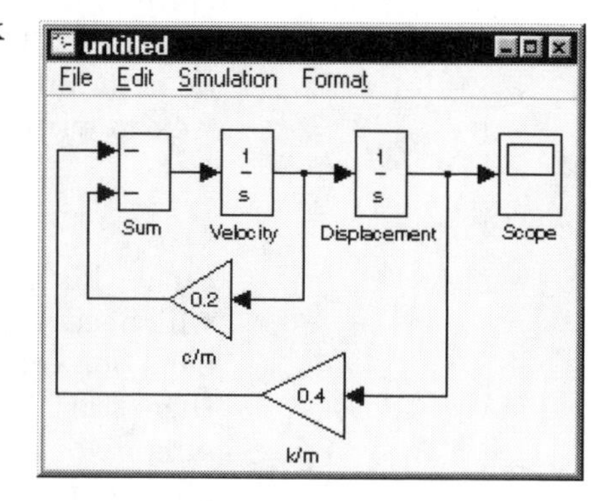

Choose **Simulation:Parameters** from the model window menu bar, and set **Stop time** to `50`. Choose **Simulation:Start** from the model window menu bar. The Scope display should be as shown in Figure 4-6 after it is autoscaled.

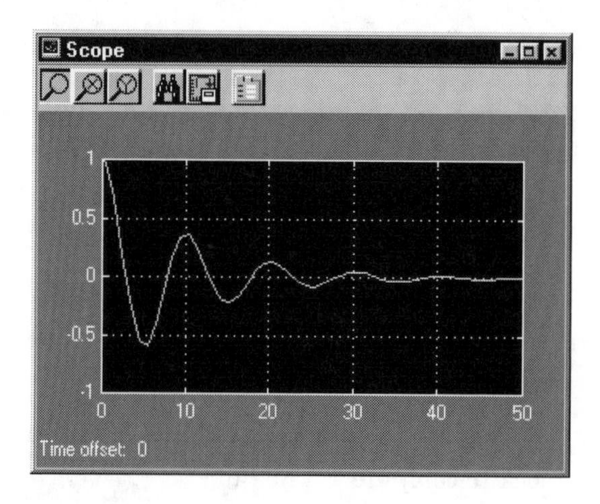

Figure 4-6 Damped second-order system response

4.2.2 Transfer Function Blocks

Transfer function notation is frequently used in control system design and system modeling. The *transfer function* can be defined as the ratio of the Laplace transform of the input to a system to the Laplace transform of the output, assuming zero initial conditions. Thus the transfer function provides a convenient input-output description of the system dynamics. As we'll see, the transfer function block is a compact notation for a composition of primitive block diagram components (Integrator, Gain, Sum, and Derivative blocks). See, for example, Lewis and Yang [3] or Ogata [4] for more information on transfer functions.

SIMULINK provides two blocks that implement transfer functions: the Transfer Fcn block and the Zero-Pole block. These blocks are equivalent, differing in the notation used to represent the transfer function. The Transfer Fcn block dialog box has two fields: **Numerator** and **Denominator**. **Numerator** contains the coefficients of the numerator of the transfer function in decreasing powers of s, and **Denominator** contain the coefficients of the denominator polynomial in decreasing powers of s. The Zero-Pole block dialog box has three fields: **Zeros**, **Poles**, and **Gain**. **Zeros** contains the zeros of the numerator of the transfer function, **Poles** contains the zeros of the denominator of the transfer function, and **Gain** scales the transfer function. Refer to the Block Browser for detailed explanations of these blocks.

Example 4-2

Consider the spring-mass-damper system depicted in Figure 4-7. This system is the same as that in Example 4-1, with the addition of a forcing function F. Assume that the system is initially at its static equilibrium point ($x = 0$, $\dot{x} = 0$) and that the forcing function is a step input of 1 lb. Ignoring friction, the equation of motion of this system is

$$m\ddot{x} + c\dot{x} + kx = F.$$

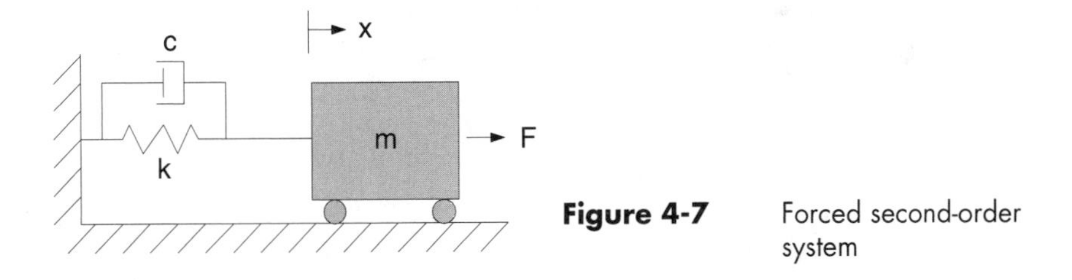

Figure 4-7 Forced second-order system

Taking the Laplace transform, and ignoring initial conditions yields

$$ms^2X(s) + csX(s) + kX(s) = F(s).$$

The ratio of the Laplace transform of the output ($X(s)$) to the Laplace transform of the input ($F(s)$) is the transfer function ($G(s)$):

$$G(s) = \frac{X(s)}{F(s)} = \frac{(1/m)}{s^2 + \frac{c}{m}s + \frac{k}{m}}.$$

Figure 4-8 shows a SIMULINK model of this system using the primitive linear blocks. Figure 4-9 shows a SIMULINK model of the same system built using a single Transfer Fcn block. The Transfer Fcn block dialog box field **Numerator** contains `[0.2]`, and **Denominator** contains `[1 0.2 0.4]`.

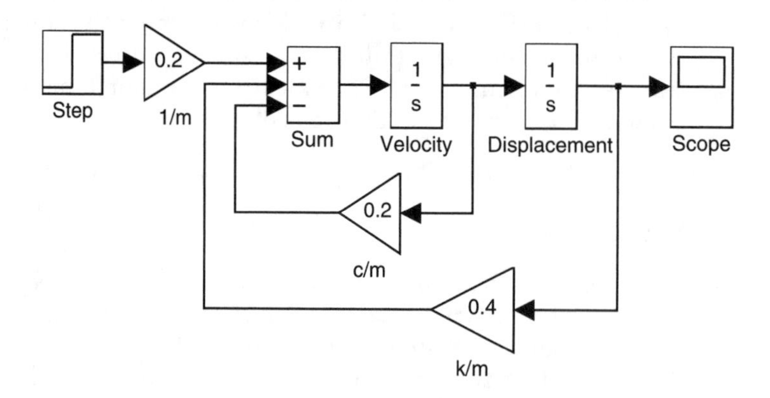

Figure 4-8 Forced second-order system with primitive blocks

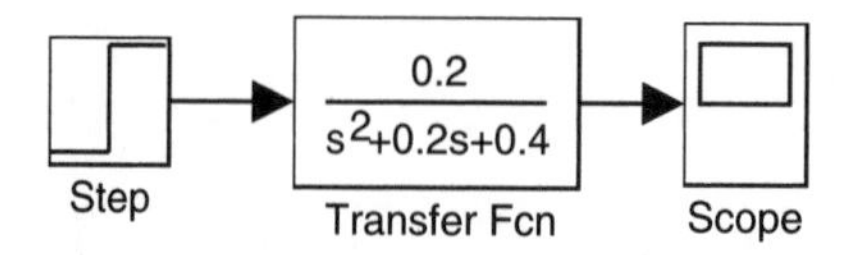

Figure 4-9 Forced second-order system using a Transfer Fcn block

4.3 Vector Linear Systems

In the examples we have examined so far, the signal lines carried scalar signals. It is often convenient to use vector signals, as they provide a more compact and easier-to-understand model. In this section we will discuss the mechanics of using vector signals and then use them with the State-Space block.

4.3.1 Vector Signal Lines

You can combine several scalar signals to form a vector signal using the Mux (multiplexer) block from the Connections block library. In Figure 4-10 we have combined three scalar signals to form a vector signal. Before you use a Mux block, you must configure it by setting the number of inputs. To make it easier to identify vector signal lines, choose **Format:Wide Vector Lines** from the model window menu bar. The components of the vector signal are referred to as `u(1)`, `u(2)`, ... , `u(n)`, where n is the number of components. The top input to the Mux block is `u(1)`, the bottom `u(n)`.

Figure 4-10 Forming a vector signal using a Mux block

The Demux block permits you to split a vector signal into a set of scalar signals. The Demux block must be configured for the correct number of outputs. Figure 4-11 illustrates splitting a vector signal into three scalar signals.

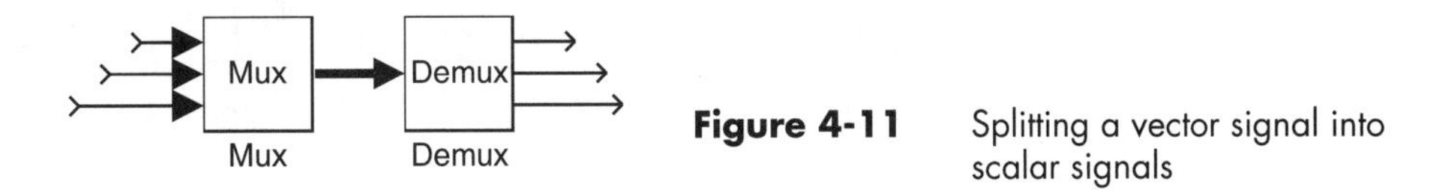

Figure 4-11 Splitting a vector signal into scalar signals

Most SIMULINK blocks will take vector inputs. The behavior of a block with a vector input will depend on both the type of block and the block configuration parameters. Linear blocks with vector inputs produce vector outputs of the same dimension as the input. The configuration parameters for linear blocks with vector inputs must be either of the same dimension as the input or scalar. In the case of scalar block parameters, SIMULINK automatically performs *scalar expansion*, which produces an implicit parameter vector of the same dimension as the input and which has all elements set to the value of the block's scalar parameter. Consider the SIMULINK model fragment illustrated in Figure 4-12. The output of the Gain block is a three-element vector in which each element is the corresponding element of the input vector multiplied by 2. Now consider Figure 4-13. Notice that the gain is the vector `[2,3,4]`. In this case, the first element of the output is the first element of the input multiplied by 2, the second element of the output is the second element of the input multiplied by 3, and the third element of the output is the third element of the input multiplied by 4. The initial value of the output of an Integrator block behaves the same as the value of gain of a Gain block with respect to vector signals.

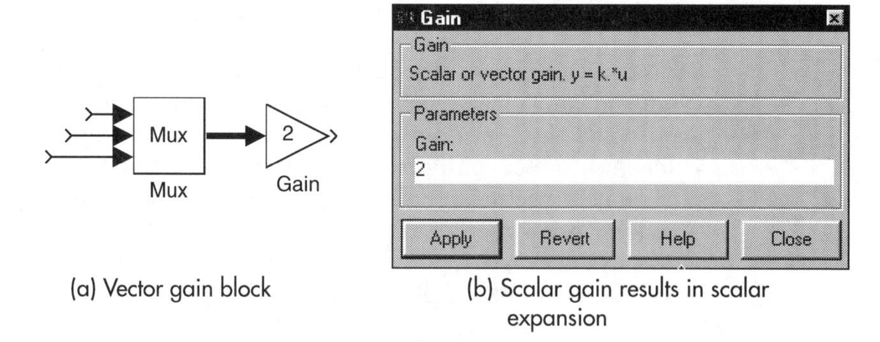

(a) Vector gain block

(b) Scalar gain results in scalar expansion

Figure 4-12 Gain block with vector input signal

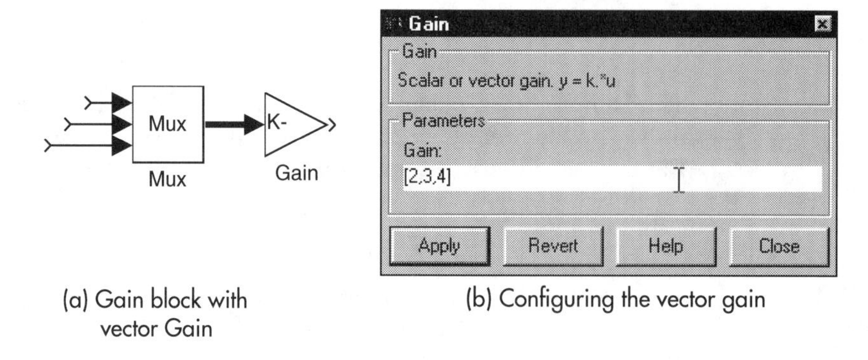

(a) Gain block with vector Gain

(b) Configuring the vector gain

Figure 4-13 Gain block with vector gain

Scalar expansion also applies to blocks with scalar inputs. For example, if the input to the Gain block in Figure 4-13 were a scalar signal, the output would be a three-element vector. The first element would be the input multiplied by 2, the second element would be the input multiplied by 3, and the third element would be the input multiplied by 4.

Other blocks, such as the Fcn block (Nonlinear block library), produce only scalar outputs regardless of the input. The Block Browser describes the behavior of each block with respect to vector signals.

4.3.2 State-Space

Before we describe the State-Space block, we will discuss the concept of state variables. A *state vector* is a set of *state variables* sufficient to describe the dynamic state of a system. The general form of the state-space model of a dynamical system is

$$\dot{\boldsymbol{x}} = \boldsymbol{f}(\boldsymbol{x}, \boldsymbol{u}, t), \tag{4-1}$$

where $\boldsymbol{x}$ is the state vector, $\boldsymbol{u}$ the *input vector,* and t time. Equation (4-1) is called the *system state equation*. We also define the system output to be

$$\boldsymbol{y} = \boldsymbol{g}(\boldsymbol{x}, \boldsymbol{u}, t). \tag{4-2}$$

Equation (4-2) is called the *output equation*. Note that we use lowercase boldface type to distinguish vectors and uppercase boldface type to distinguish matrixes.

The so-called *natural state variables* are simple quantities such as position, velocity, temperature, or electrical current. For a mechanical system, the natural state variables are positions and velocities. For electrical circuits, the natural state variables could be voltages or currents. The natural state variables are not the only set of state variables we can choose. In fact, any independent linear combination of a valid set of state variables is also a valid set of state variables. Example 4-3 illustrates the selection of state variables for a single-degree-of-freedom system.

Example 4-3

Consider the pendulum of Figure 4-14.

Figure 4-14 Pendulum model

The dynamic state of the pendulum is defined by its position and velocity. A natural set of state variables is the deflection angle and the rate of change of deflection:

$$x_1 = \theta$$

$$x_2 = \dot{\theta}.$$

The state variable approach is particularly useful in modeling linear systems because we can take advantage of matrix notation to describe very complex systems in a compact form. Additionally, we can compute the system response using matrix arithmetic.

Consider the spring-mass-damper system of Figure 4-7. This is a second-order system with one degree of freedom, so we must choose two state variables. Choosing

$$\begin{aligned} x_1 &= x \\ x_2 &= \dot{x} \end{aligned} \tag{4-3}$$

the time rates of change of the two state variables are

$$\begin{aligned} \dot{x}_1 &= x_2 \\ \dot{x}_2 &= -\frac{k}{m}x_1 - \frac{c}{m}x_2 + \frac{1}{m}F \end{aligned} \tag{4-4}$$

These equations can be written in matrix notation:

$$\dot{\boldsymbol{x}} = \boldsymbol{A}\boldsymbol{x} + \boldsymbol{B}\boldsymbol{u} \tag{4-5}$$

where

$$\boldsymbol{x} = \begin{bmatrix} x_1 \\ x_2 \end{bmatrix} \tag{4-6}$$

$$\boldsymbol{A} = \begin{bmatrix} 0 & 1 \\ -\frac{k}{m} & -\frac{c}{m} \end{bmatrix} \tag{4-7}$$

$$\boldsymbol{B} = \begin{bmatrix} 0 \\ \frac{1}{m} \end{bmatrix} \tag{4-8}$$

$$u = F \tag{4-9}$$

Matrix $\boldsymbol{A}$ is frequently called the *system matrix*. The system matrix is always square. Matrix $\boldsymbol{B}$ is the *input matrix*. The number of rows in the input matrix is the same as the number of state variables, and the number of columns is the same as the number of inputs. Equation (4-5) is the matrix form of the system state equation for a linear system. Note that the system and input matrixes are not characteristics of the system. Different choices of state variables will result in different system and input matrixes.

The state variables we have defined may be considered internal states of a system. The state variables are not necessarily the system outputs. The outputs may consist of a subset of the states, or may consist of a linear combination of the system states and the inputs. The output equation for a linear system is

$$y = \boldsymbol{C}\boldsymbol{x} + \boldsymbol{D}\boldsymbol{u} \tag{4-10}$$

If we choose the output of the system depicted in Figure 4-7 to be the position of the mass (x), referring to Equation (4-3),

$$y = x_1 \tag{4-11}$$

we have

$$\boldsymbol{C} = \begin{bmatrix} 1 & 0 \end{bmatrix} \tag{4-12}$$

and

$$\boldsymbol{D} = 0 \tag{4-13}$$

$\boldsymbol{C}$ is called the *output matrix*. $\boldsymbol{D}$ is called the *direct transmittance matrix* because if it is non-zero, the input is transmitted directly to the output. For more information on the state-space concept, refer to Kuo [2] or Ogata [4].

4.3.3 State-Space Block

The State-Space block implements a linear state-space model of a system or a portion of a system. The block dialog box has fields for each of the four linear state-space matrixes (***A***, ***B***, ***C***, ***D*** as defined in Section 4.3.2), and a fifth field **Initial conditions**. Each field contains a MATLAB matrix.

Example 4-4

Consider again the spring-mass system shown in Figure 4-7. In this example we will model the impulse response of the system. Substituting in the model parameters, equation (4-4) is

$$\begin{aligned} \dot{x}_1 &= x_2 \\ \dot{x}_2 &= -0.4x_1 - 0.2x_2 + 0.2\delta(t), \end{aligned}$$

where $\delta(t)$ is the unit impulse function. The system matrix is

$$\boldsymbol{A} = \begin{bmatrix} 0 & 1 \\ -0.4 & -0.2 \end{bmatrix},$$

and the input matrix is

$$\boldsymbol{B} = \begin{bmatrix} 0 \\ 0.2 \end{bmatrix}.$$

Define the output to be the block position,

$$\boldsymbol{C} = \begin{bmatrix} 1 & 0 \end{bmatrix}.$$

There is no direct transmittance, so we can also set

$$\boldsymbol{D} = 0.$$

We can approximate the unit impulse as a positive step function followed by a negative step:

$$\delta(t) \cong 100u(t) - 100u(t - 0.01),$$

as was illustrated in Example 3-1.

Figure 4-15 shows a SIMULINK model of this formulation of the equations of motion and the State-Space block dialog box. Note that the State-Space block includes initial conditions for each state variable.

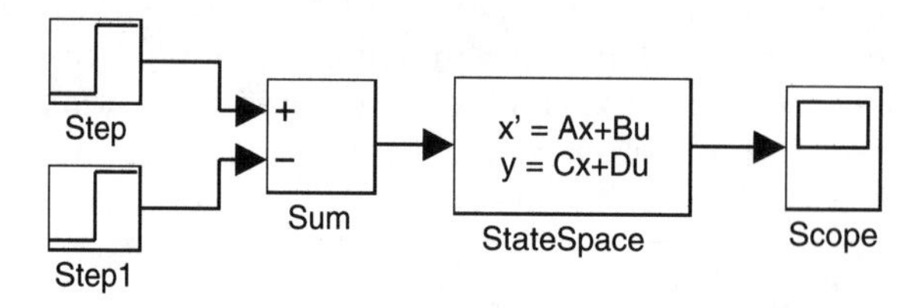

Figure 4-15 State-space model of spring-mass system

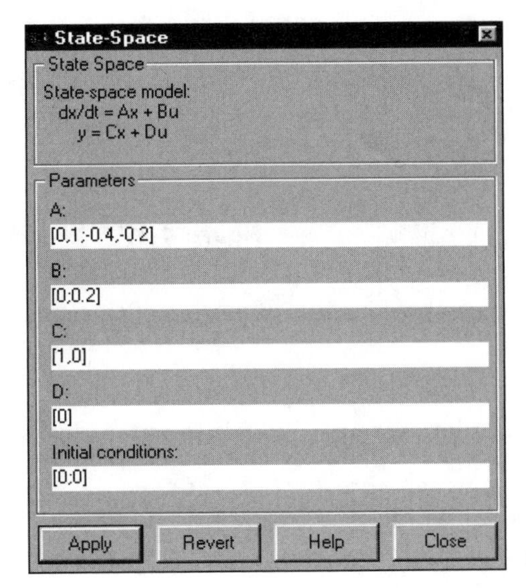

Figure 4-16 State-Space dialog box

4.4 Modeling Nonlinear Systems

SIMULINK provides a variety of blocks for modeling nonlinear systems. These blocks are in the Nonlinear block library. The behavior of nonlinear blocks with respect to vector inputs varies. Some blocks, such as the relay, produce vector outputs of the same dimension as the input. Other blocks produce only scalar outputs, or scalar or vector outputs depending on the dimensionality of the inputs. Consult the Block Browser for details on a particular block.

Example 4-5

To illustrate the use of several blocks from the Nonlinear block library, we will model the motion of the rocket-powered cart shown in Figure 4-17. The cart is powered by two opposing rocket motors. The controller fires the left motor if the sum of cart velocity and displacement is negative, and fires the right motor if the sum of cart velocity and displacement is positive. The objective of the controller is to bring the cart to rest at the origin. This type of control is sometimes called *bang-bang control*. (This is actually a simple example of a very powerful technique called *sliding mode control*, explained by Khalil [1].)

Assume the cart mass is 5 slugs and that the motor force F is 1 lbf.

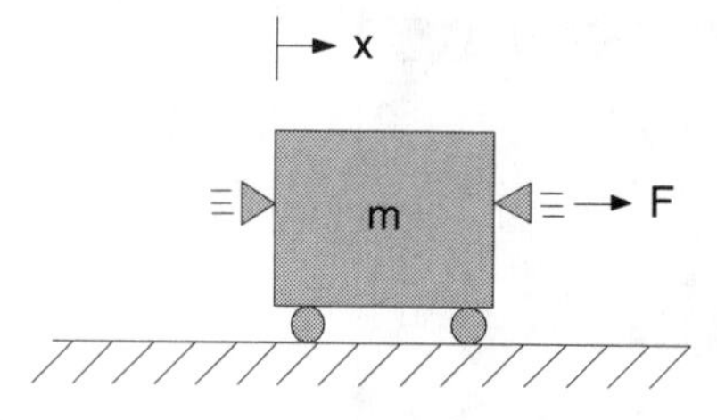

Figure 4-17 Rocket-powered cart

We begin by writing the equation of motion of the cart:

$$m\ddot{x} = F$$

This is a second-order system, so two Integrator blocks are needed to solve for the cart position.

Open a new model window and copy two Integrator blocks from the Linear block library. Label the blocks as shown.

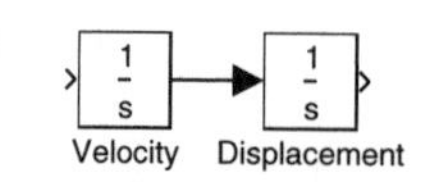

The input to the first Integrator block is acceleration. Solving the equation of motion for acceleration,

$$\ddot{x} = \frac{F}{m}.$$

Add a Gain block to multiply the rocket motor force by $1/m$. The input to this Gain block will be the motor force F.

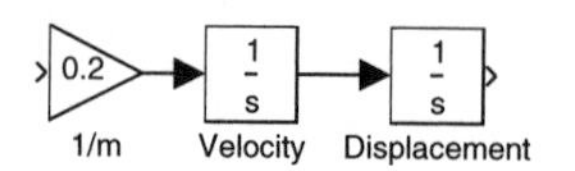

The motor force F is 1.0 if the sum of velocity and displacement is negative, and –1.0 if the sum is positive. We can build a suitable bang-bang controller using a Sum block from the Linear block library and a Sign block from the Nonlinear block library. The output of a Sign block is 1.0 if its input is positive, –1.0 if its input is negative, and 0.0 if its input is exactly 0.

Add the Sum block and Sign block and connect them as shown. The Sum block is configured with two minus signs (– –) because the motor force is to be opposite in sign to the sum of velocity and displacement.

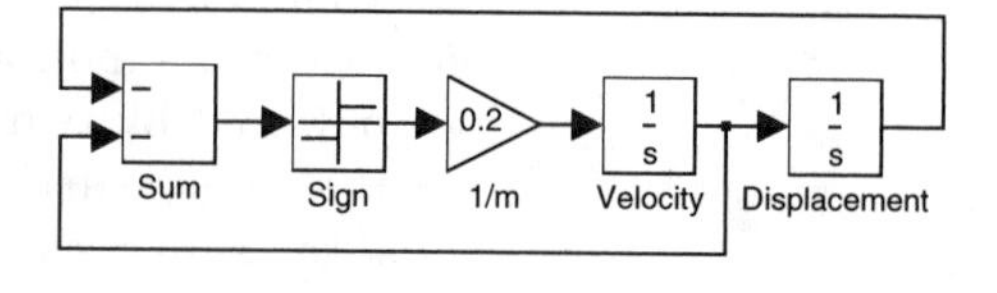

To display the results of the simulation of the model, we will use an XY Graph block to draw a *phase plot* as the simulation progresses. A phase plot is a plot of velocity versus displacement.

Drag an XY Graph block from the Sinks block library. Connect the output of the displacement Integrator to the X input (the upper input port) and the output of the velocity integrator to the Y (lower) input port.

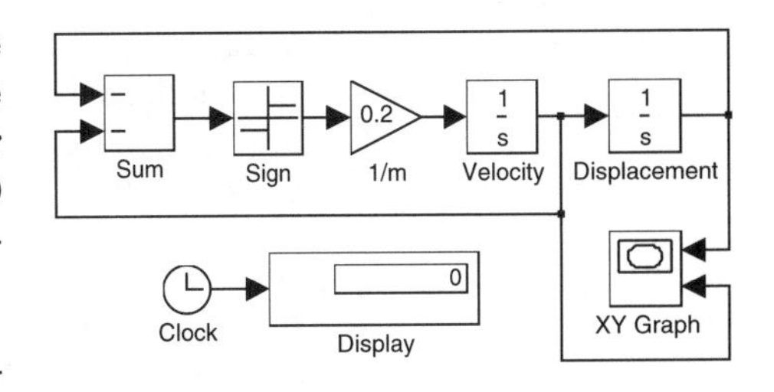

To display simulation time as the simulation progresses, add a Clock block from the Sources block library, and connect it to a Display block from the Sinks block library.

We don't know how long it will take the solution to reach the origin, so for convenience, let's add logic to the model to cause the simulation to stop automatically when the objective is reached. Specifically, we will add logic that causes the simulation to stop when the sum of the absolute values of velocity and displacement falls below a threshold value of 0.01. We can accomplish this task with a Stop Simulation block from the Sinks block library. This block forces SIMULINK to stop the simulation when the block's input is nonzero. We wish for the input to the Stop Simulation block to be zero until

$$|x| + |\dot{x}| \leq 0.01.$$

Copy two Abs blocks from the Nonlinear block library and a Sum block from the Linear block library. Next copy a Relational Operator block from the Nonlinear block library. Double click on the Relational Operator block and choose <= from the **Operator** drop-down list. Add a Constant block from the Sources block library and set its value to `0.01`. Finally, drag a Stop Simulation block from the Sinks block library, and connect the signal lines as shown.

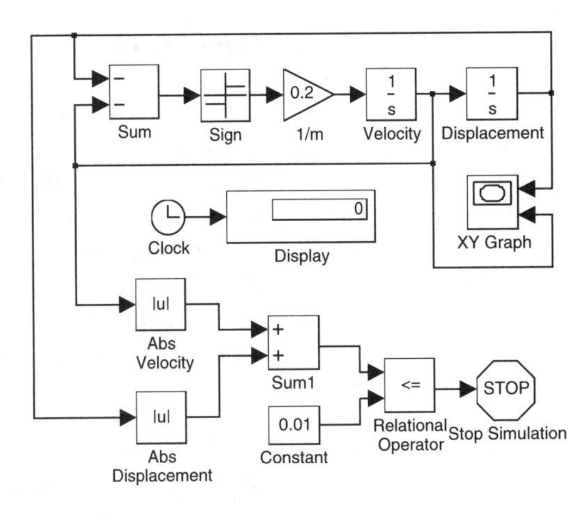

We will assume that the cart is initially at rest, displaced 1 ft to the right. So set the **Initial condition** for the velocity Integrator block to `0`, and for the displacement Integrator block to `1`.

The Sign block switches instantaneously when its input changes sign. However, any physically realizable switch takes a finite amount of time to change state. To model that behavior, we can use a fixed-step-size solver. Let's assume that the switch time is 0.05 seconds. Choose **Simulation:Parameters** from the model window menu bar. Set the Solver options **Type** fields to Fixed-step and ODE5 (Dormand-Prince). Then set the **Fixed step size** to `0.05`. Set **Stop time** to `200`.

Now we are ready to run the simulation. Choose **Simulation:Start** from the model window menu bar. The simulation will execute and will stop when the cart is at rest at the origin. The XY Graph should appear as shown in Figure 4-18.

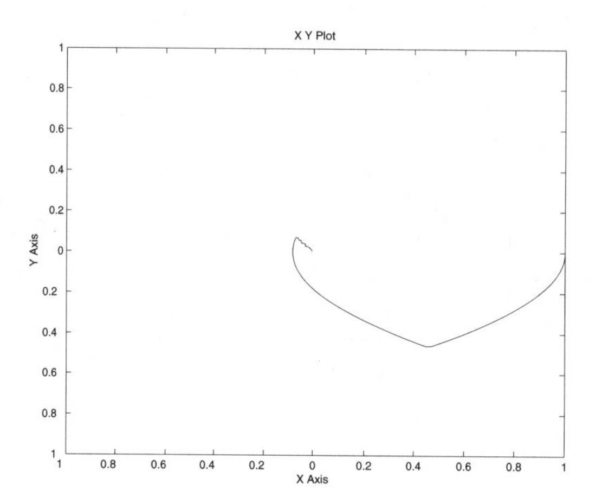

Figure 4-18 Cart phase plot

If you look closely at the phase plot, you will see that as the phase trajectory approaches the origin, it oscillates about the line $\dot{x} = -x$. This phenomenon is called *chatter* (discussed in detail by Khalil [1]). Try replacing the Sign block with a Saturation block with **Upper limit** set to `0.05` and **Lower limit** set to `–0.05`. Also try replacing the Sign block with a Dead Zone block, with **Start of dead zone** set to `–0.05` and **End of dead zone** set to `0.05`. When you use the Saturation block or Dead Zone block instead of the Sign block, you can use a variable-step solver since neither block changes instantaneously from –1 to +1.

4.4.1 Function Blocks

Two particularly useful nonlinear blocks are the Fcn (C function) block and the MATLAB Fcn block. Both blocks perform specified mathematical operations on the input, but the blocks have some important differences. SIMULINK evaluates the Fcn block much faster than the MATLAB Fcn block. However, the Fcn block can't perform matrix computations, use MATLAB functions, or produce a vector output. Because of the speed advantage, it is always better to use the Fcn block unless the special capabilities of the MATLAB Fcn block are needed.

Fcn Block

The Fcn block is shown in Figure 4-19. The dialog box contains a single field that contains an expression in the C language syntax. The expression operates

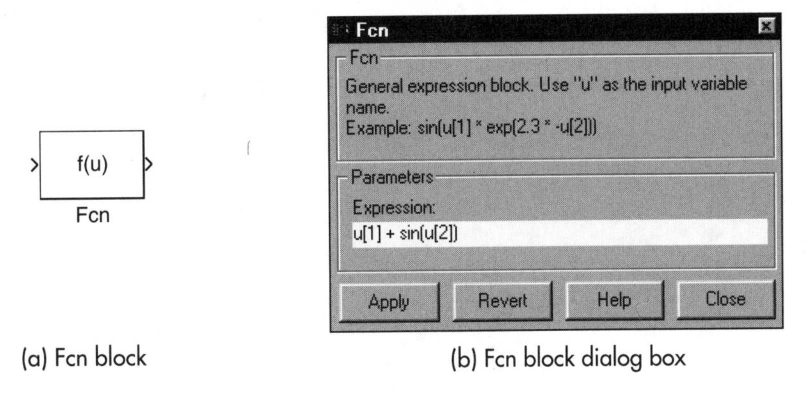

(a) Fcn block (b) Fcn block dialog box

Figure 4-19 Fcn block

on the elements of the block input vector, referring to the elements as u[*n*], where *n* is the desired element. If the input is a scalar, the input is referred to as u[1]. Note the use of square brackets ("[" and "]"), as in the C language, rather than parentheses, as in the MATLAB language. The block may also use as a parameter any variable currently (at the time of execution of the simulation) defined in the MATLAB workspace. If the workspace variable is a scalar, it may be referred to by its name. For example, if there is a scalar workspace variable a and two inputs to the Fcn block, a valid expression would be a*(sin(u[1])+u[2]). If the workspace variable is a vector or matrix, it is referred to using the appropriate MATLAB (not C) syntax A(1), or A(2,3) (note the parentheses here). The Fcn block can perform all the standard scalar mathematical functions (such as sin, abs, atan), C syntax relational operations (==, !=, >, <, >=, <=), and C syntax logical operations (&& (logical AND), || (logical OR)). The Fcn block produces a scalar output.

MATLAB Fcn Block

The MATLAB Fcn block is more powerful than the Fcn block in that it can perform matrix computations and produce vector outputs. However, it is also much slower than the Fcn block, and thus should only be used in situations in which the Fcn block can't be used. The MATLAB Fcn block is illustrated in Figure 4-20. The dialog box has two fields. The first may contain any valid MATLAB expression, in standard MATLAB syntax. The value of the expression is the block output. The *n*th element of the block input vector is referred to as u(n), similar to the Fcn block. If the function field contains a MATLAB function

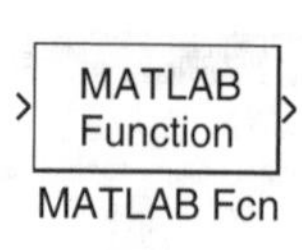

(a) MATLAB Fcn block

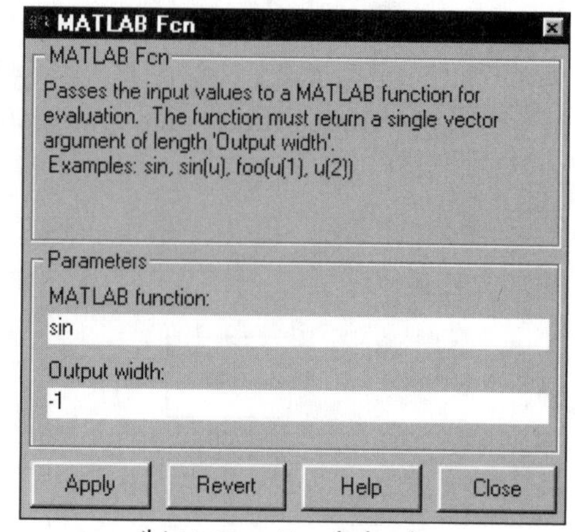

(b) MATLAB Fcn dialog box

Figure 4-20 MATLAB Fcn block

with no arguments (as in Figure 4-20b), the operation is performed on all elements of the input. The second input field specifies the dimension of the output vector. Enter –1 if the output is to be the same width as the input. No matter if **Output width** is specified explicitly or allowed to default to the input width: The vector dimension of the result of the expression in the **MATLAB function** field must be the same as the value specified in **Output width**.

Example 4-6 will illustrate the use of nonlinear blocks to build a simple model of a car and proportional gain cruise control. The model will include aerodynamic drag, the gravity force due to climbing hills, and wind.

Example 4-6

Consider the automobile traveling on a straight hilly road shown in Figure 4-21.

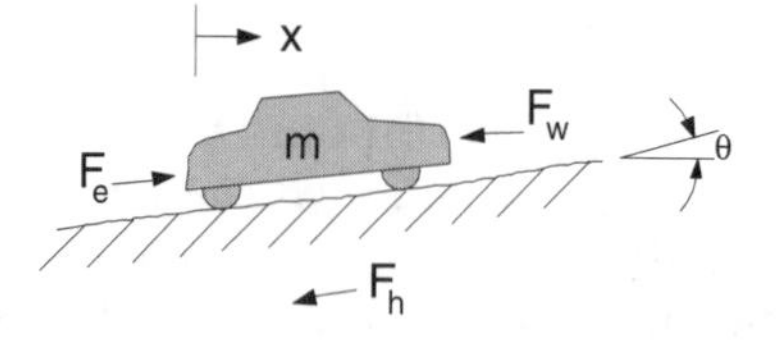

Figure 4-21 Automobile on a hilly road

There are three forces acting on the automobile: the forward thrust produced by the engine and transmitted through the tires (or the braking force if negative) (F_e), the aerodynamic force (including wind) (F_w), and the tangential component of gravity as the automobile climbs and descends hills (F_h). Applying Newton's second law, the equation of motion of the automobile can be written

$$m\ddot{x} = F_e - F_w - F_h,$$

where m represents the mass of the automobile and x the distance traveled. F_e must have upper and lower bounds. The upper bound is the maximum force that the engine can transmit through the wheels to the road, and the lower bound is the maximum braking force. We will assume that $-2000 \le F_e \le 1000$, with units of pounds, and that the mass is 100 slugs.

The aerodynamic force is the product of the drag coefficient (C_D), the automobile's frontal area (A), and the dynamic pressure (P), where

$$P = \frac{\rho V^2}{2}$$

and ρ represents the air density and V the sum of the automobile speed and wind speed (V_w). Assume that

$$\frac{C_D A \rho}{2} = 0.001$$

and that the wind speed varies sinusoidally with time according to the rule

$$V_w = 20\ \sin(0.01t)$$

so that the aerodynamic force can be approximated by

$$F_w = 0.001(\dot{x} + 20\sin(0.01t))^2$$

Next assume that the road angle varies sinusoidally with distance according to the rule

$$\theta = 0.0093\sin(0.0001x)$$

Then the hill force is

$$F_h = 30\sin(0.0001x)$$

We will control the automobile speed using the simple proportional control law

$$F_c = K_e(\dot{x}_{desired} - \dot{x})$$

Here F_c is the commanded engine (or braking) force, $\dot{x}_{desired}$ is the commanded speed (ft/s), and K_e is the feedback gain. Thus the commanded engine force is proportional to the speed error. The actual engine force (F_e) is, as stated earlier, bounded from above by the maximum engine thrust and from below by the maximum braking force. We choose $K_e = 50$.

A SIMULINK model of this system is shown in Figure 4-22. We will simulate the motion of the car for 1000 s.

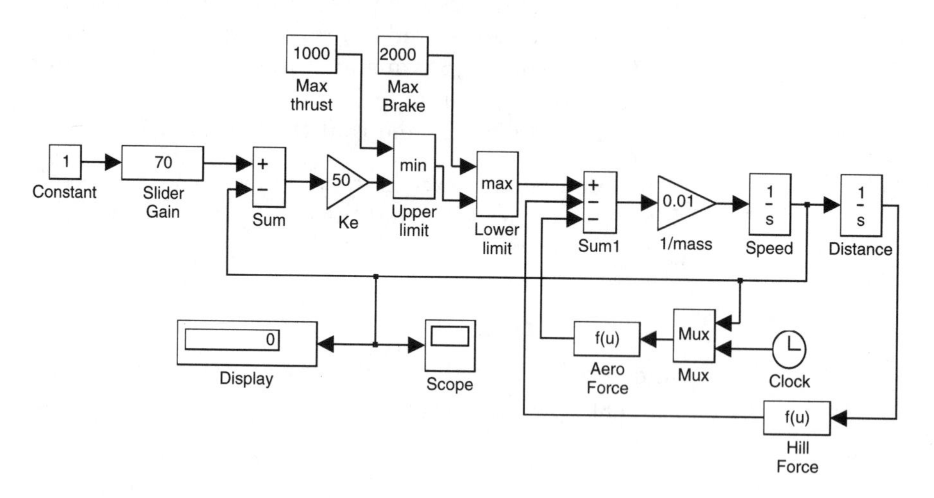

Figure 4-22 Automobile model with proportional speed control

The input to the proportional controller is the desired automobile speed in ft/s. This is implemented with a Slider Gain block (from the Linear block library) with a constant input. Double click on the Slider Gain block to open a slider window, which allows you to vary the desired speed while the simulation runs.

The proportional controller consists of a Sum block, which computes the speed error (the difference between the commanded speed and the actual speed), and a Gain block.

The upper and lower limits on engine force are imposed using MinMax blocks. The constant blocks labeled `Max thrust` and `Max Brake` together with the Min and Max blocks are used here to illustrate the use of those blocks. This part of the model could be replaced with a Saturation block from the Nonlinear block library. (Why don't you try that?)

The nonlinear hill and aerodynamic forces are computed by Fcn blocks. The **Expression** field of the block dialog box for the Fcn block labeled `Aero Force` contains `0.001*(u[1]+20*sin(0.01*u[2]))^2`. The **Expression** field for the Fcn block labeled `Hill Force` contains `30*sin(0.0001*u[1])`.

A Display block serves as a speedometer (which reads ft/s), and the speed is plotted using a Scope block.

This model is a good example of a slightly stiff system. To observe the effect of this stiffness, run the model using solver ODE45 and then repeat the simulation using ODE15S. See Chapter 8 for a discussion of stiff systems.

4.5 Summary

In this chapter we have discussed the use of SIMULINK to model continuous systems. We started with scalar, linear, time-invariant systems and progressed to vector linear systems using the state-space concept. Finally, we discussed modeling nonlinear continuous systems.

4.6 References

1. Khalil, Hassan K., *Nonlinear Systems*, 2nd ed. Upper Saddle River, N.J.: Prentice Hall, 1996. An excellent text on the analysis and design of nonlinear systems.
2. Kuo, Benjamin C., *Automatic Control Systems.* Englewood Cliffs, N.J.: Prentice Hall, 1995, pp 226–230. This is a comprehensive text covering all the basics of control system analysis and design.
3. Lewis, Paul H., and Yang, Charles, *Basic Control Systems Engineering.* Upper Saddle River, N.J.: Prentice Hall, 1997. This text provides an introduction to control systems analysis and design and includes a brief introduction to SIMULINK. SIMULINK is used in many of the examples.
4. Ogata, Katsuhiko, *Modern Control Engineering.* Englewood Cliffs, N.J.: Prentice Hall, 1990. This book presents a thorough coverage of the standard techniques for the analysis and design of controls for continuous systems.
5. Scheinerman, Edward C., *Invitation to Dynamical Systems.* Upper Saddle River, N.J.: Prentice Hall, 1996. This text provides a good introduction to nonlinear systems.
6. Shahian, Bahram, and Hassul, Michael, *Control System Design Using MATLAB.* Englewood Cliffs, N.J.: Prentice Hall, 1993. This book provides an introduction to MATLAB programming and uses MATLAB to solve many of the standard problems in classical control and modern control theory.
7. Strum, Robert D., and Kirk, Donald E., *Contemporary Linear Systems Using MATLAB.* Boston, Mass., PWS Publishing Co., 1994. Chapter 2 provides a nice introduction to continuous systems and the state-space concept.
8. Vidyasagar, M., *Nonlinear Systems Analysis*, 2nd ed. Englewood Cliffs, N.J.: Prentice Hall, 1993. This book presents a detailed coverage of the analysis of nonlinear systems.

5

Discrete-Time Systems

In this chapter we will discuss using SIMULINK to model discrete-time systems. We'll start with a brief introduction to discrete-time systems. Then we'll discuss modeling discrete-time systems using blocks from the Linear, Discrete, and Nonlinear block libraries. Finally, we'll discuss modeling hybrid systems, which include both continuous and discrete components.

5.1 Introduction

SIMULINK provides extensive capabilities for modeling discrete-time systems. The Discrete block library contains simple discrete blocks, such as Zero- and First-Order Holds, Discrete Integrators, and the Unit Delay. The Nonlinear block library contains additional blocks that are useful in discrete-time systems, such as logical operators and a Combinational Logic (truth table) block. Additionally, there are a number of blocks in the Linear (such as Sum and Gain) and Nonlinear block libraries (for example, Product, Sign, Fcn) that have the same purpose when used with continuous or discrete-time models. Discrete-time models may be single rate or they may be multirate, including offset sample times. SIMULINK will optionally color code the signal lines in multirate models such that the signal line color indicates the sample period.

5.1.1 Discrete-Time System Overview

A discrete system is a system that may be represented using difference equations and that operates on discrete signals. A discrete signal can be represented as a sequence of pulses, as shown in Figure 5-1.

A discrete system takes as its input one or more discrete signals and produces one or more discrete signals as its output. A discrete-time system is a discrete system that may change only at specific instants of time (see Ogata [3]). In the majority of discrete-time control systems, the signals are not inherently discrete. In these systems the discrete signals are extracted from continuous signals by a process known as *sampling*. Figure 5-2 illustrates the sampling process.

The type of sampling illustrated in Figure 5-2 is performed using two devices: a *sampler* and a *zero-order hold*. The sampler periodically closes a switch for an

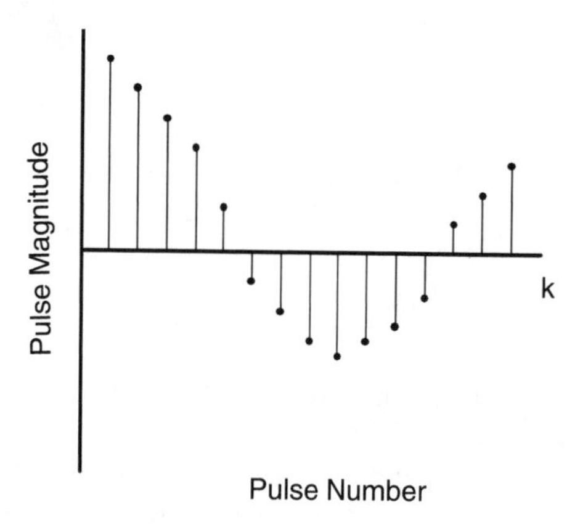

Figure 5-1 Discrete signal

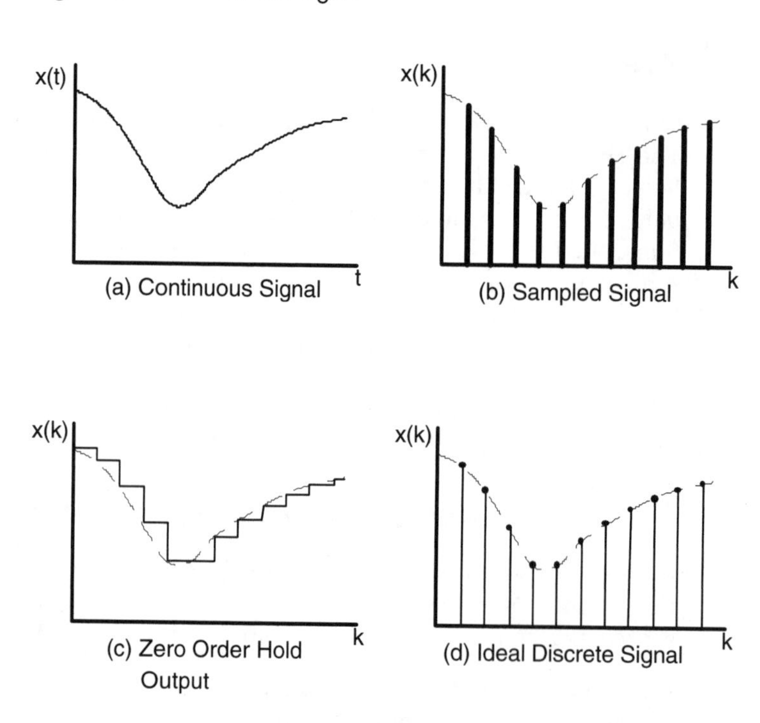

Figure 5-2 Sampling a continuous signal to produce a discrete signal

instant. Each time the switch is momentarily closed, a pulse of very short (theoretically zero) duration, and equal in magnitude to the input signal, is produced (Figure 5-2b). The spacing between the pulses is the *sampling period* (T). The zero-order hold follows the sampler and clamps its output at the last value of its input, producing a stairstep signal as shown in Figure 5-2c. A discrete controller requires a sequence of numbers as its input. The *analog to digital (A/D) converter* is a device that converts the stairstep signal of Figure 5-2c into a sequence of numbers, as represented by the pulse sequence shown in Figure 5-2d. The combination of the sampler and zero-order hold block is illustrated in Figure 5-3 along with the SIMULINK equivalent.

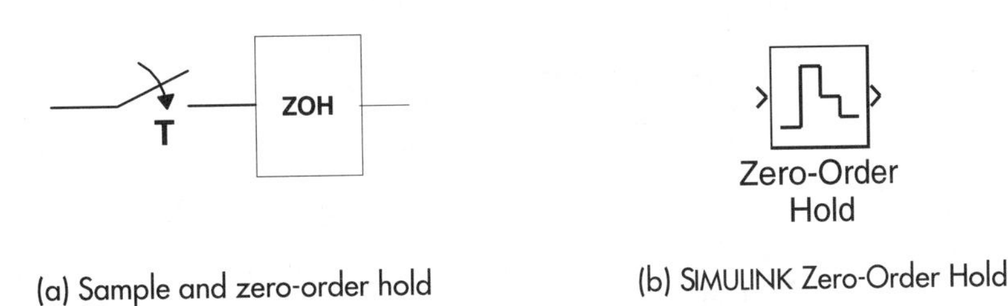

(a) Sample and zero-order hold

(b) SIMULINK Zero-Order Hold

Figure 5-3 Sample and zero-order hold

We refer to signals that vary with time using the notation $x(t)$. The corresponding notation for discrete signals is $x(k)$, where k is the ordinal pulse number. The mapping from sample space to continuous time is

$$x(t) = x(kT), \tag{5-1}$$

where T is the sampling period. See Ogata [3] or Phillips and Nagle [5] for a detailed discussion of sampling and discrete systems.

5.2 Scalar Linear Discrete-Time Systems

Modeling scalar linear discrete-time systems is very similar to modeling continuous systems. Discrete-time models can use the Gain and Sum blocks from the Linear block library. These blocks behave the same in discrete systems as they do in continuous systems. The Discrete block library contains the discrete analogs to the continuous Integrator and Transfer Fcn blocks. We will discuss these blocks next and show how they may be used.

Each discrete block is assumed to have a sampler at its input and a zero-order hold at its output. Discrete blocks have the additional configuration parameter **Sample time**. **Sample time** is either a scalar interval between samples, or a two-element vector consisting of the interval between samples and an offset or time skew. For example, if the block is to have a sample interval of 1.5 seconds and no offset, **Sample time** would be set to `1.5`. If the sample time is to be 0.75 seconds with a 0.25 second offset, the **Sample time** would be set to `[0.75, 0.25]`.

All of the differential equation solvers listed in the **Simulation:Parameters Solver options** are compatible with discrete systems. A special solver, named "discrete (no continuous states)," is the best choice for purely discrete systems because this solver is optimized for these systems.

5.2.1 Unit Delay

The Unit Delay is the fundamental discrete-time block. The unit delay is sometimes called a *shift register* or *time-delay element*. The output of the unit delay block is the input at the previous sample time. The unit delay represents the difference equation

$$y(k) = x(k-1) \tag{5-2}$$

where y is the output sequence and x is the input sequence. The Unit Delay block dialog box has two fields. The first field is **Initial condition**. This field contains the value of the block output at the start of the simulation. The second field is **Sample time**.

Example 5-1

In this example we will model the amortization of an automobile loan. At the end of each month, the loan balance is the sum of the balance at the beginning of the month and the interest for the month, less end of the month payment. If we represent the balance at the end of the month as $b(k)$, the balance at the end of month k is

$$b(k) = rb(k-1) - p(k)$$

where $r = 1 + i$ and i is the monthly interest rate.

Assume that the initial loan balance is $15,000, the interest rate is 1% per month (12% annual interest), and the monthly payment is $200. Compute the loan balance after 100 payments.

Figure 5-4 shows a SIMULINK model of this system. The Unit Delay block computes $b(k-1)$. The Unit Delay block **Initial condition** is the initial loan balance (15,000). The Unit Delay block **Sample time** is set to 1. Set solver type to **Fixed-step, discrete (no continuous states)**, and let **Start time** be 0 and **Stop time** be 100. After running the simulation, the Display block shows the ending balance.

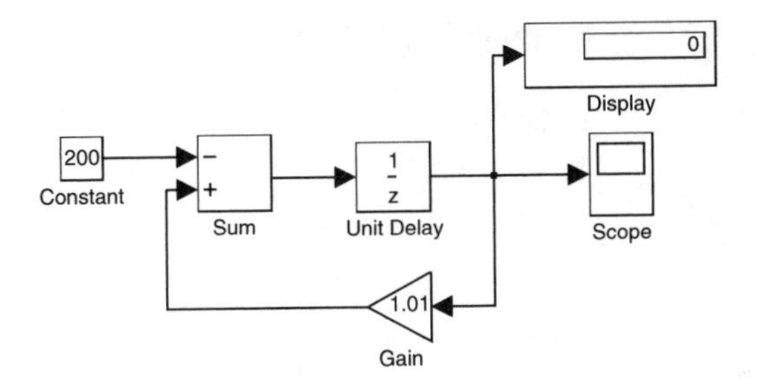

Figure 5-4 Block diagram of loan amortization

5.2.2 Discrete-Time Integrator

The output of the Discrete-Time Integrator block is an approximation of the time integral of the input signal. That is, it is a discrete approximation of a continuous integrator. Thus the output of the Discrete-Time Integrator block approximates

$$y(k) = y(k-1) + \int_{T(k-1)}^{Tk} u(t)dt, \tag{5-3}$$

where $u(t)$ is the input to the integrator, $y(k)$ is the output, and T is the sample period.

The Discrete-Time Integrator block dialog box is shown in Figure 5-5. The fields in this dialog box are the same as the fields for the Integrator dialog box with two additions. For a detailed discussion of the Integrator block dialog box fields, refer to Section 4.2.1. The two additional fields are **Integrator method** and **Sample time**.

Integrator method is one of three choices: **Forward Euler**, **Backward Euler**, and **Trapezoidal**. Because the Discrete-Time Integrator is a discrete block, it has a sample and zero-order hold at its input. Thus at each time step, a Discrete-Time Integrator block has access to only two values of $u(t)$: $u(Tk)$ at its input, and $u(T(k-1))$ via a built-in unit delay. Each of the three choices for **Integrator method** approximates $u(t)$ differently. We will briefly discuss each method.

Note that you will sometimes see the definitions of forward and backward Euler integration reversed from SIMULINK's convention. For example, Kuo [1] refers to the method SIMULINK calls forward Euler integration as backward rectangular integration, and the method SIMULINK calls backward Euler integration as forward rectangular integration.

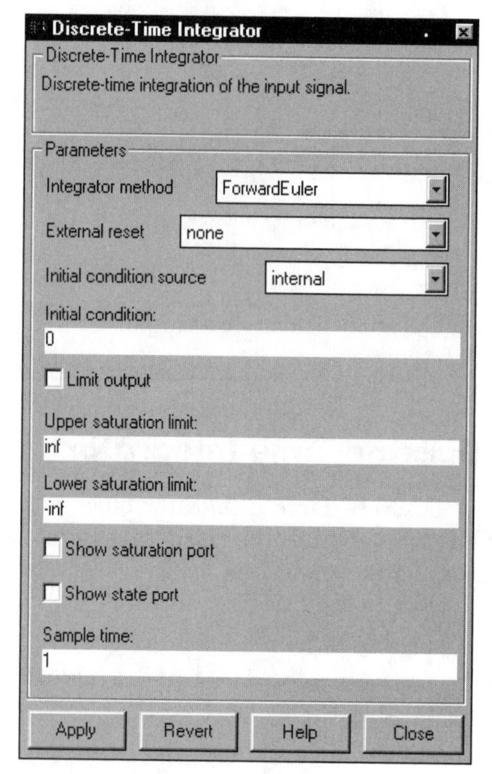

Figure 5-5 Discrete-Time Integrator block dialog box

Forward Euler

Forward Euler integration is based on the approximation $u(t) = u(T(k-1))$. Thus forward Euler integration approximates

$$y(k) = y(k-1) + Tu(T(k-1)) \tag{5-4}$$

Taking the Z transform of Equation (5-4),

$$Y(z) = z^{-1}Y(z) + Tz^{-1}U(z) \tag{5-5}$$

Rearranging terms, the Z transfer function of the forward Euler integrator is

$$\frac{Y(z)}{U(z)} = \frac{T}{z-1} \tag{5-6}$$

This is graphically depicted in Figure 5-6. The icon for the Discrete-Time Integrator block displays the transfer function.

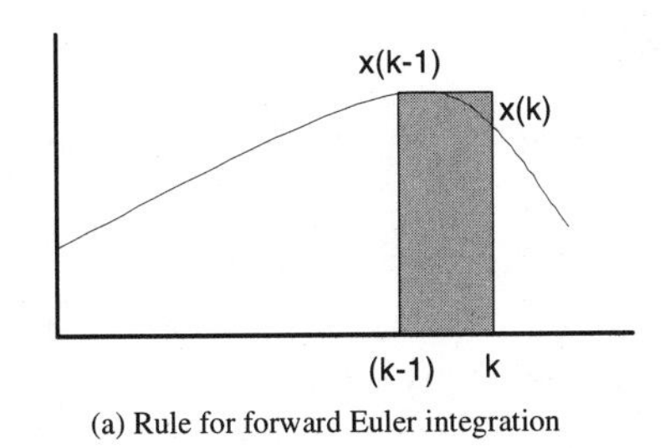

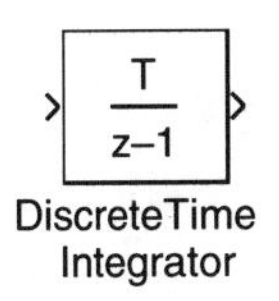

(a) Rule for forward Euler integration

(b) Discrete-Time Integrator block configured to do forward Euler integration

Figure 5-6 Forward Euler integrator

Backward Euler

Backward Euler integration is based on the approximation $u(t) = u(Tk)$. Thus backward Euler integration approximates

$$y(k) = y(k-1) + Tu(Tk) \tag{5-7}$$

The corresponding Z transfer function is

$$\frac{Y(z)}{U(z)} = \frac{Tz}{z-1} \tag{5-8}$$

This is graphically depicted in Figure 5-7 along with the backward Euler version of the Discrete-Time Integrator block.

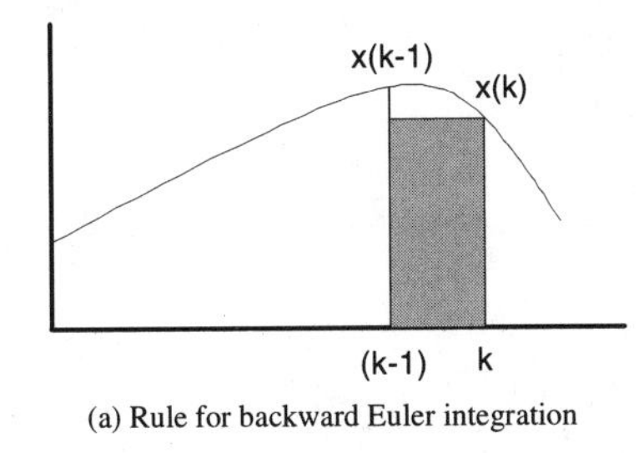

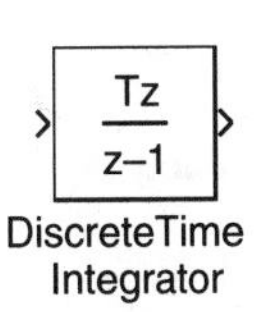

(a) Rule for backward Euler integration

(b) Discrete-Time Integrator block configured to do backward Euler integration

Figure 5-7 Backward Euler integrator

Trapezoidal Integration

Trapezoidal integration is based on the approximation

$$u(t) = \frac{u(Tk) + u(T(k-1))}{2} \tag{5-9}$$

Thus trapezoidal integration approximates

$$y(k) = y(k-1) + \frac{Tu(Tk) + Tu(T(k-1))}{2}. \tag{5-10}$$

The corresponding Z transfer function is

$$\frac{Y(z)}{U(z)} = \frac{T(z+1)}{2(z-1)}. \tag{5-11}$$

This is graphically depicted in Figure 5-8.

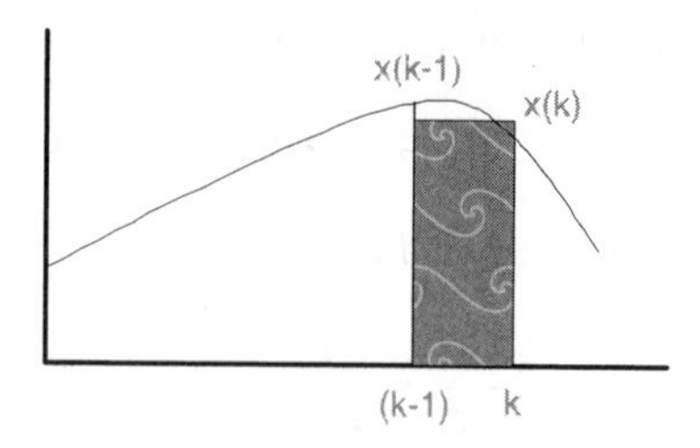

(a) Rule for trapezoidal integration

$\frac{T(z+1)}{2(z-1)}$

Discrete-Time Integrator

(b) Discrete-Time Integrator block configured to do trapezoidal integration

Figure 5-8 Trapezoidal integrator

Example 5-2

Consider the automobile loan problem of Example 5-1. The loan balance at the end of the month can be computed as

$$b(k) = b(k-1) + \int_{T(k-1)}^{Tk} (ib(k-1) - p)dt$$

Examining the integrand, we can see that this problem can be solved using forward Euler integration. Figure 5-9 shows the SIMULINK model revised to use a Discrete-Time Integrator block with **Integrator method** set to **Forward**

Euler. The Discrete-Time Integrator block **Initial condition** is set to `15,000`, and Sample time to `1`. Note that in this case, the feedback gain is set to `0.01`.

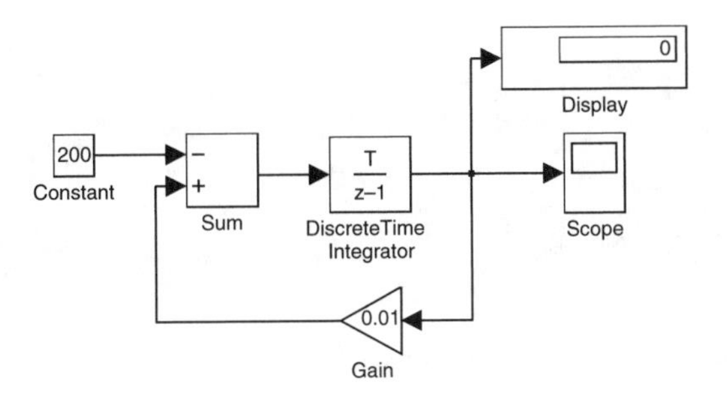

Figure 5-9 Loan amortization using a Discrete-Time Integrator block

5.2.3 Discrete Transfer Function Blocks

A discrete transfer function is analogous to a continuous transfer function. A discrete transfer function is defined as the ratio of the Z transform of the input to a system to the Z transform of the output. The Discrete block library provides three blocks that implement discrete transfer functions: Discrete Filter, Discrete Transfer Fcn, and Discrete Zero-Pole. These blocks are equivalent, differing only in the definitions of the coefficients for the numerator and denominator polynomials. The Discrete Filter block requires vectors of coefficients of polynomials of ascending powers of z^{-1}, whereas the Discrete Transfer Fcn block requires vectors of coefficients of polynomials of descending powers of z. Thus these blocks are identical, differing only in the way the transfer function is displayed on the block icon. The Discrete Zero-Pole block requires vectors of zeros (numerator) and poles (denominator) of the transfer function and a gain that scales the transfer function.

Example 5-3

Control systems frequently contain filters to remove high frequency noise from input signals. The Student Edition of MATLAB includes the Signal Processing Toolbox, which provides a variety of filter design algorithms. You can include a filter designed using MATLAB in a SIMULINK model using a Discrete Transfer Fcn block.

Suppose we are designing a control system with a noisy sinusoidal input, and we need to filter out the noise. The sampling period is 0.1 second, and we wish to remove signals with a frequency higher than 0.2 Hz. Design a fourth-order Butterworth filter using the MATLAB command `butter` as shown in Figure 5-10. (See Orfanidis [4] to learn about digital filters.)

```
EDU» [B,A]=butter(4,0.2)
B =
    0.0048    0.0193    0.0289    0.0193    0.0048
A =
    1.0000   -2.3695    2.3140   -1.0547    0.1874
```

Figure 5-10 Designing a Butterworth filter

We can test the filter with the SIMULINK model shown in Figure 5-11. The top Sine Wave block is configured to a frequency of 0.5 rad/s and an amplitude of 1. The lower Sine Wave block is configured to a frequency of 10 rad/s and an amplitude of 0.4, and represents the unwanted high-frequency noise. The Discrete Transfer Fcn block field **Numerator** contains the vector `[0.0048 0.0193 0.0289 0.0193 0.0048]` and **Denominator** contains `[1.0000 -2.3695 2.3140 -1.0547 0.1874]`. **Sample time** contains `0.1`. The model is configured to use the discrete solver since there are no continuous states. **Stop time** is set to 20. The Mux block creates a vector signal containing the unfiltered input and the filtered output of the Discrete Transfer Fcn block. Scope block field **Save data to workspace** is checked. After running the simulation, the MATLAB commands shown in Figure 5-12 are used to plot the filtered and unfiltered output signal, resulting in the plots shown in Figure 5-13.

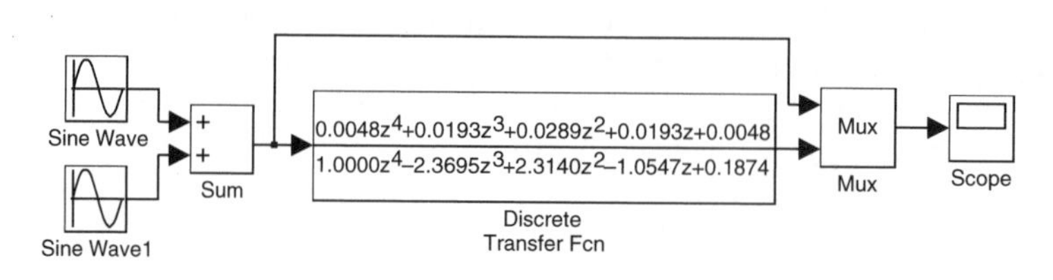

Figure 5-11 SIMULINK model to test filter

```
EDU» t = ScopeData(:,1) ;
EDU» y_raw = ScopeData(:,2) ;
EDU» y_filt = ScopeData(:,3) ;
EDU» subplot(2,1,1)
EDU» plot(t,y_raw);
EDU» title('Unfiltered Signal')
EDU» grid
EDU» subplot(2,1,2)
EDU» plot(t,y_filt);
EDU» title('Filtered Signal')
EDU» xlabel('Time (sec)')
EDU» grid
```

Figure 5-12 MATLAB commands to plot filtered signal

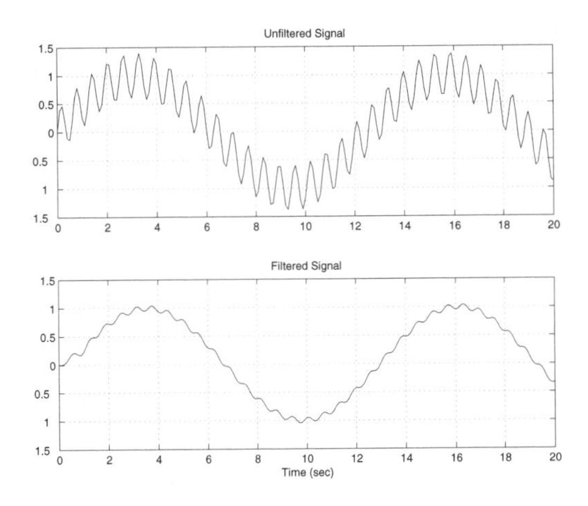

Figure 5-13 Filtering a noisy signal

5.3 Logical Blocks

The Logical Operator and Combinatorial Logic blocks found in the Nonlinear block library are also useful in modeling discrete-time systems. These are discrete-valued rather than discrete-time blocks since they have neither samplers at their inputs nor zero-order holds at their outputs. Although these blocks can be used in continuous models, they are more commonly used in discrete-time models. The Logical Operator block can be configured to perform any of the following logical operations: AND, OR, NAND, NOR, XOR, NOT. The Logical Operator block can be configured to accept any number of inputs. If the input signals are vectors, each must be of the same size, and the output will also be a vector.

The Combinatorial Logic block implements a truth table. The input to the Combinatorial Logic block is a vector of n elements. The truth table must have 2^n rows, arranged such that the values of a row's inputs provide an index into the table. The first element in the input vector is the leftmost column in this binary index. Thus, if there is a single input, there will be two rows, the first corresponding to an input of 0, the second corresponding to an input of 1 (or any nonzero number). If there are two inputs, there must be four rows in the table. The block parameter, **Truth table**, consists of one or more column vectors, in which each row represents the block output corresponding to the row's index. Each column in **Truth table** produces a different output, and thus a single Combinatorial Logic block can implement multiple logical operations. The block outputs do not have to be 0 or 1; any number is acceptable.

Example 5-4

We will build Combinatorial Logic blocks that implement the following three expressions, and compare these three blocks with equivalent models built using Logical Operator blocks.

Expression	Equivalent
$\bar{a}$	NOT a
ab	a AND b
$ab + bc$	(a AND b) OR (b AND c)

The truth table for $\bar{a}$ is shown in Figure 5-14a. The corresponding **Truth table** parameter is the output column, entered as `[1;0]`. A SIMULINK model that implements this expression using both Logical Operator blocks and a Combinatorial Logic block is shown in Figure 5-14b.

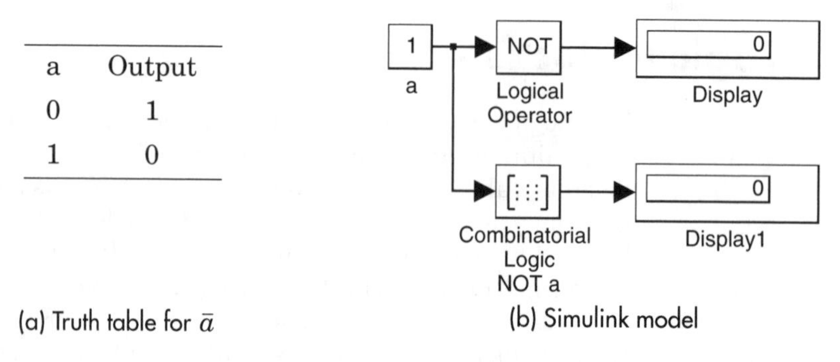

a	Output
0	1
1	0

(a) Truth table for $\bar{a}$

(b) Simulink model

Figure 5-14 NOT expression

The truth table for ab is shown in Figure 5-15a. Note that we have chosen a to be the leftmost column; thus the input vector to the block must be $[a,b]$. The block **Truth table** parameter is the same as the output column, `[0;0;0;1]`. Figure 5-15b shows a SIMULINK model that implements ab using both Logical Operator blocks and a Combinatorial Logic block. The input vector is composed using a Mux block such that the components are in the order `[a,b]`.

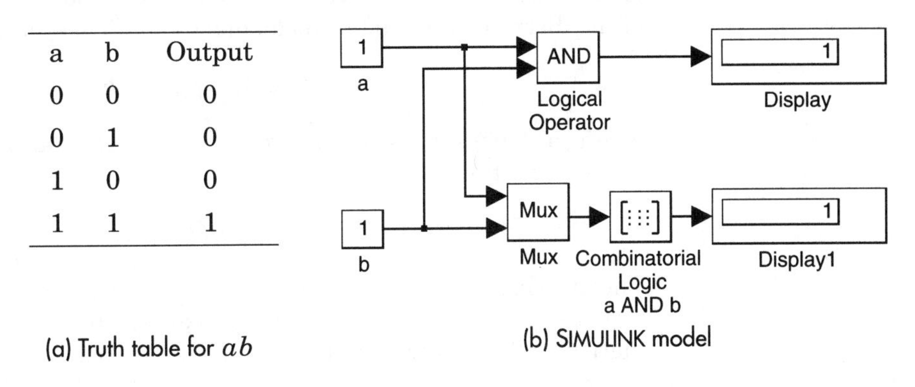

a	b	Output
0	0	0
0	1	0
1	0	0
1	1	1

(a) Truth table for ab

(b) SIMULINK model

Figure 5-15 AND expression

The final combinatorial logic expression, $ab + bc$, has the truth table shown in Figure 5-16a. Examining the output column, the **Truth table** parameter is `[0;0;0;1;0;0;1;1]`. Figure 5-16b implements the expression $ab + bc$ using both Logical Operator blocks and a Combinatorial Logic block.

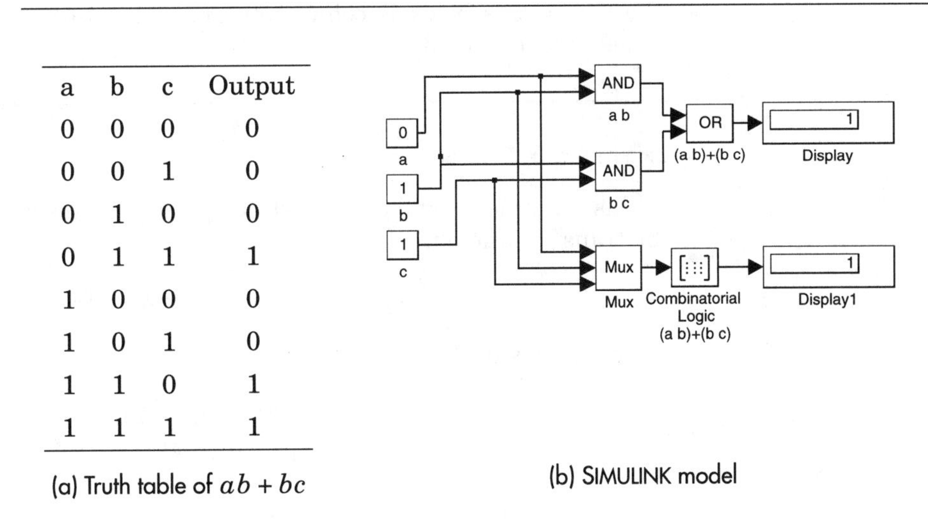

a	b	c	Output
0	0	0	0
0	0	1	0
0	1	0	0
0	1	1	1
1	0	0	0
1	0	1	0
1	1	0	1
1	1	1	1

(a) Truth table of $ab + bc$

(b) SIMULINK model

Figure 5-16 Model of the expression $ab + bc$

5.4 Vector Discrete-time Systems

The state-space concept discussed in Section 4.3.2 is also useful with discrete-time systems. Whereas the state variables in continuous systems represent derivatives or linear combinations of derivatives, the state variables in discrete systems represent portions of sequences or linear combinations of portions of sequences. Thus the discrete equivalent to Equation (4-1) is the general form of the discrete state-space model of a dynamical system:

$$\boldsymbol{x}(k+1) = \boldsymbol{f}(\boldsymbol{x}, \boldsymbol{u}, \boldsymbol{k}) \tag{5-12}$$

where, as before, $\boldsymbol{x}$ is the state vector and $\boldsymbol{u}$ the input vector. The corresponding output equation is

$$\boldsymbol{y}(k) = \boldsymbol{g}(\boldsymbol{x}, \boldsymbol{u}, \boldsymbol{k}) \tag{5-13}$$

The matrix notation for linear discrete systems is similar to the model for continuous systems. The same four matrixes (system matrix, input matrix, output matrix, and direct transmittance matrix) are used. Thus the system equation is

$$x(k+1) = \boldsymbol{A}\boldsymbol{x}(k) + \boldsymbol{B}\boldsymbol{u}(k) \tag{5-14}$$

The output equation is

$$\boldsymbol{y}(k) = \boldsymbol{C}\boldsymbol{x}(k) + \boldsymbol{D}\boldsymbol{u}(k) \tag{5-15}$$

Once a discrete state-space model is derived, it may be implemented in SIMULINK using the Discrete State-Space block found in the Discrete block library. The use of this block is completely analogous to the State-Space block discussed in Section 4.3.3.

Example 5-5

In this example we will develop state-space matrixes for a system described by the transfer function

$$y(k+1) = y(k) - 2y(k-1) + u(k)$$

Define state variables

$$\begin{aligned} x_1(k) &= y(k-1) \\ x_2(k) &= y(k) \end{aligned}$$

The corresponding state equations are

$$\begin{aligned} x_1(k+1) &= x_2(k) \\ x_2(k+1) &= x_2(k) - 2x_1(k) + u(k) \end{aligned}$$

In matrix form we have

$$\boldsymbol{x}(k+1) = \begin{bmatrix} 0 & 1 \\ -2 & 1 \end{bmatrix} \boldsymbol{x}(k) + \begin{bmatrix} 0 \\ 1 \end{bmatrix} u$$

so the system matrix is

$$\boldsymbol{A} = \begin{bmatrix} 0 & 1 \\ -2 & 1 \end{bmatrix}$$

and the input matrix is

$$\boldsymbol{B} = \begin{bmatrix} 0 \\ 1 \end{bmatrix}$$

There is no direct transmittance, so $\boldsymbol{D}$ is zero. The output variable is $y(k)$, so

$$\boldsymbol{C} = \begin{bmatrix} 0 & 1 \end{bmatrix}$$

5.5 Multirate Discrete-time Systems

Many discrete-time systems involve subsystems that operate at different rates, and with phase differences as well. For example, a typical computer has numerous discrete subsystems, including the central processing unit, serial and parallel interface controllers, disk drive and video controllers, and input devices, such as the keyboard and mouse. Models of communications systems and transaction processes also consist of subsystems that operate at different rates. Modeling multirate discrete-time systems with SIMULINK is similar to modeling single rate discrete-time systems. However, multirate systems require careful attention to sample times and offsets.

SIMULINK's sample time colors capability provides help in keeping track of sample times. This capability will automatically color code blocks and signal lines corresponding to up to five different values of sample time. To activate this feature, choose **Format:Sample Time Colors**. If you change the model after activating sample time colors, choose **Edit:Update Diagram** to update the colors. Table 5-1 lists the sample time colors and their meanings.

Table 5-1 Sample time colors

Color	Meaning
Black	Continuous blocks
Magenta	Constant blocks (used with SIMULINK Real-Time Workshop)
Yellow	Hybrid (groups of blocks with varying sample times or mixed continuous and discrete elements)
Red	Fastest discrete sample time
Green	Second fastest discrete sample time
Blue	Third fastest discrete sample time
Light Blue	Fourth fastest discrete sample time
Dark Green	Fifth fastest discrete sample time
Cyan	Triggered sample time (used with Triggered subsystems. See Chapter 6.)

Inherently continuous blocks (such as Integrators) and inherently discrete blocks (such as Unit Delays) are assigned colors using the definitions in Table 5-1. Signal lines and blocks that are neither inherently continuous nor discrete (such as Gain blocks) are assigned colors based on the sample times of their inputs. Thus, if a signal line is carrying the output of an Integrator, it will be black, and if it is carrying the output of a Unit Delay, it will be colored according to the sample time of the Unit Delay. If the input to a block (such as a Sum block) consists of several signals such that the sample times of all signals are integer multiples of the sample time of the fastest signal, the block is colored corresponding to the fastest signal. If the sample times are not integer multiples of the fastest signal, the block is colored black.

Only the five fastest discrete sample times are assigned unique colors. If there are more than five sample times, all blocks sampled slower than the fifth fastest sample time are colored yellow.

Note that SIMULINK does not provide a mechanism to help identify phase relationships among blocks. If there are portions of a model with the same sample time but different values for sample offset (phase), you must keep track of the phase manually.

Example 5-6

Control systems for discrete processes frequently operate at a lower frequency than the update frequency of the process, usually due to limitations of computer speed. Additionally, display systems usually are updated at a frequency sufficiently low that the display is readable. As a simple example of a multirate system, suppose that some process to be controlled behaves according to the following discrete state-space equations:

$$x_1(k+1) = x_1(k) + 0.1x_2(k)$$

$$x_2(k+1) = -0.05\sin x_1(k) + 0.094x_2(k) + u(k)$$

where $u(k)$ is the input. The process is assumed to have a sample time of 0.1 second. We will control the process using proportional control with a sample time of 0.25 second, and update the display every 0.5 second. A SIMULINK model of this system appears in Figure 5-17. The model is annotated to indicate the colors of the signal lines and blocks.

The blocks in the section of the model with the red signal lines are configured with a sample time of 0.1 second. This section represents the process dynamics.

The section with green signal lines models the controller. This is a proportional controller. The Zero-Order Hold block causes the controller to update at its sample interval of 0.25 second. The controller produces an output signal that is proportional to the difference between the setpoint (here 0.75) and the value of the input to the Zero-Order Hold (x_1) at the most recent sample time.

The section with the blue signal lines models the display device, here a Scope block. The Zero-Order Hold in this section is configured with a sample time of 0.5 second.

Running the simulation results in the trajectory shown in Figure 5-18.

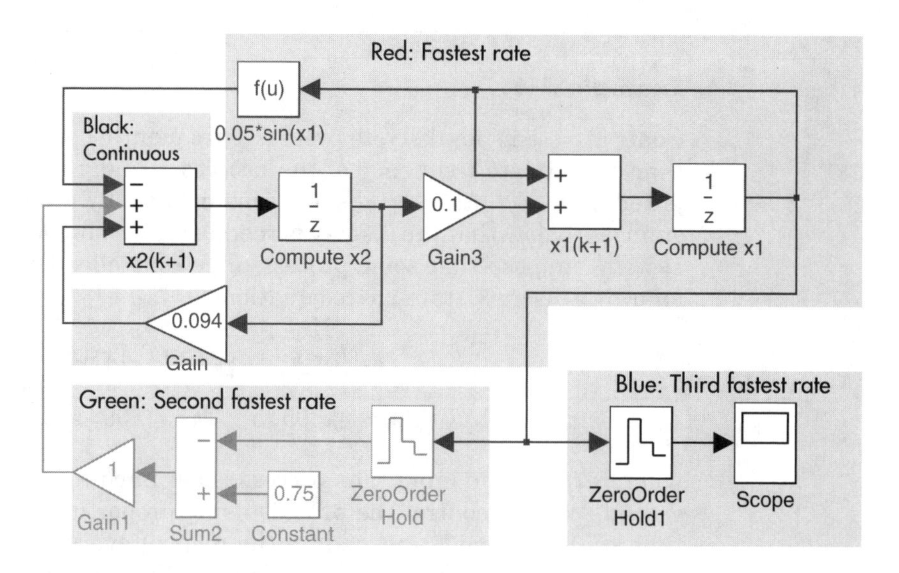

Figure 5-17 Multirate system

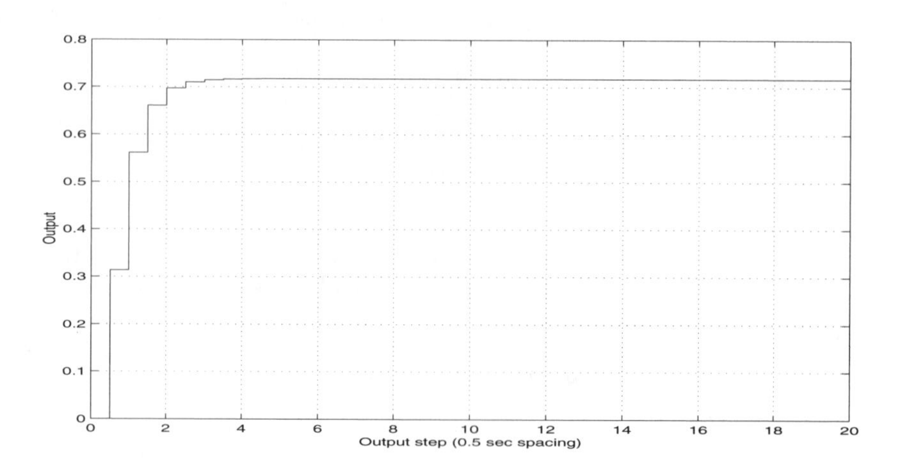

Figure 5-18 Multirate system output

5.6 Hybrid Systems

Hybrid systems consist of both discrete and continuous components. The most common hybrid system consists of a continuous physical process controlled using discrete logic or a computer. Modeling hybrid systems in SIMULINK is straightforward, as illustrated in the following example.

Example 5-7

To illustrate the construction of a hybrid model, let's replace the continuous controller in Example 4-6 with a discrete proportional-integral-derivative (PID) controller with a sample time of 0.5 second.

Figure 5-19 illustrates a continuous PID controller. The controller consists of three sections, each of which operates on the difference (v) between the plant output and the commanded value of the plant output. The proportional section produces a signal proportional to the difference between the commanded value of the system output and the actual value. Thus the output of the proportional section is

$$u_p = K_p v$$

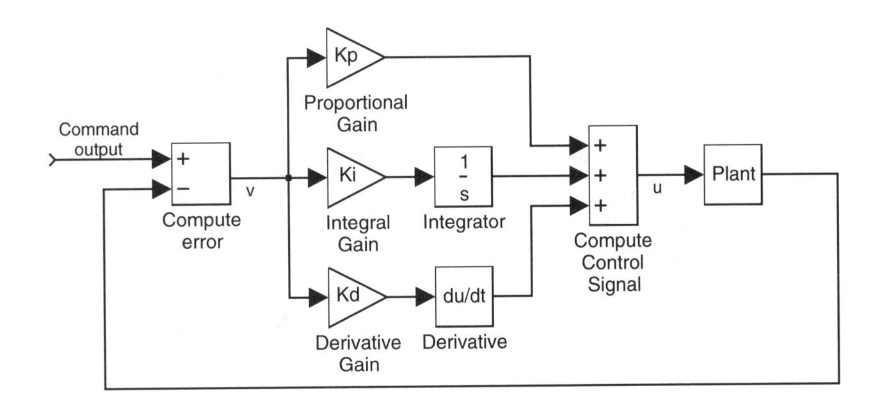

Figure 5-19 Continuous PID controller

The integral section of a PID controller is intended to reduce the steady state error. This component produces an output that is proportional to the time integral of the error signal:

$$u_i = K_i \int_0^t v dt$$

One problem with the integral section is that if the plant's response to changes in the input signal (u) is relatively slow, u_i can grow rapidly. This phenomenon is called *wind-up*. Wind-up can be avoided by placing upper and lower bounds on the value of u_i.

The derivative section of a PID controller provides damping. Its output is proportional to the rate of change of v:

$$u_d = K_d \dot{v}$$

A discrete PID controller replaces the integral section with a discrete integrator and the derivative section with a discrete approximation to a derivative block. For a detailed discussion of discrete PID controllers, refer to Ogata [3]. A first-order numerical derivative approximation is

$$u_d(k) \approx \frac{v(k) - v(k-1)}{T}$$

The transfer function of this derivative approximation is

$$\frac{U_d(z)}{V(z)} = \frac{K_d}{T}\left(\frac{z-1}{z}\right)$$

Figure 5-20 shows a SIMULINK model of an automobile using a discrete PID controller with $K_p = 50$, $K_i = 0.75$, and $K_d = 75$. The model is identical to the model in Example 4-6, except for the controller. The proportional part of the controller consists of a Zero-Order Hold and a proportional Gain block. The proportional gain (50) in this controller is the same as the proportional gain in the continuous proportional controller in Example 4-6. The integral part of the PID controller consists of a Discrete-Time Integrator block and a Gain block (set to 0.75). In the Discrete-Time Integrator block, we selected **Limit output**, and set the saturation limits to ± 100 to control wind-up. The derivative part of the controller is constructed using a Discrete Transfer Fcn block and a Gain

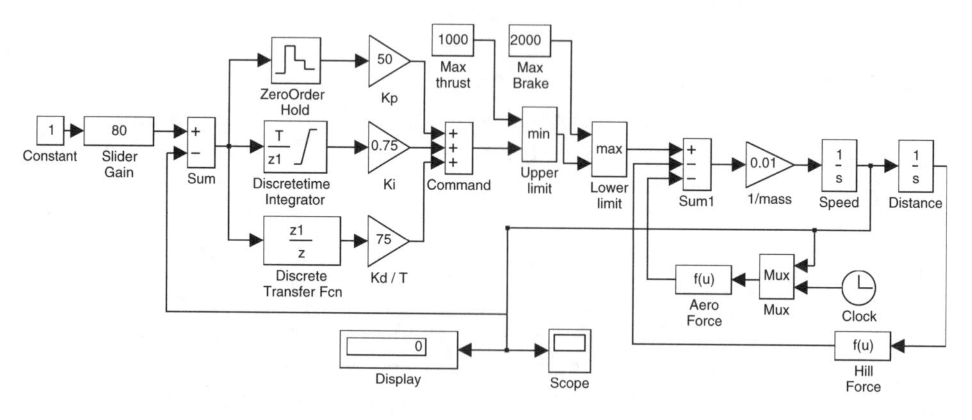

Figure 5-20 Car model with discrete controller

block (set to 75). In this example the simulation was configured to run for 1000 seconds, and the slider gain was set to command a speed of 80 ft/s. The Scope display is shown in Figure 5-21.

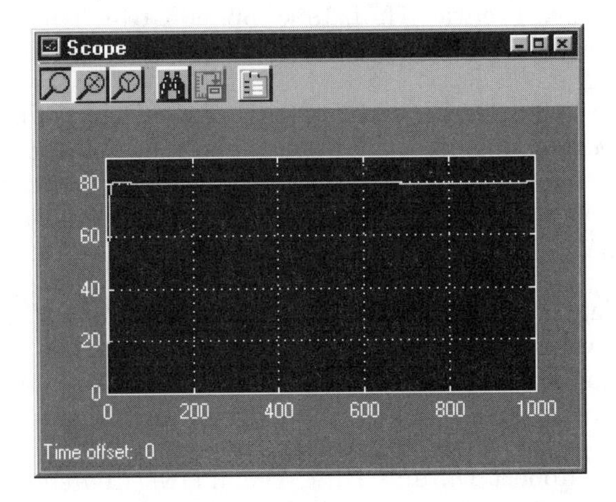

Figure 5-21 Car speed controller performance

5.7 Summary

In this chapter, we have discussed the use of SIMULINK to model discrete-time systems. We started with scalar linear discrete-time systems, discussing the Unit Delay, Discrete-Time Integrator, and Discrete Transfer Fcn blocks in particular. We briefly discussed vector discrete systems. We concluded with discussions of multirate discrete-time systems and hybrid systems.

5.8 References

1 Kuo, Benjamin C., *Automatic Control Systems, 7th Ed.* Englewood Cliffs, N.J.: Prentice Hall, 1995, pp. 839–840. This text covers both continuous and discrete control systems design and includes lots of practical examples, including MATLAB scripts. Although SIMULINK is not used, the block diagrams are easily converted to SIMULINK models.

2 Ogata, Katsuhiko, *Designing Linear Control Systems with MATLAB*, Englewood Cliffs, N.J.: Prentice Hall, 1993, pp. 50–67. This book presents brief tutorials and MATLAB implementations of several important linear systems design techniques, including pole placement, state observers, and linear quadratic regulators.

3 Ogata, Katsuhiko, *Discrete-Time Control Systems*, 2nd ed. Englewood Cliffs, N.J.: Prentice Hall, 1994, pp. 114–118.

4 Orfanidis, Sophocles J., *Introduction to Signal Processing*. Englewood Cliffs, N.J.: Prentice Hall, 1996, pp. 605–614. This book provides detailed coverage of Z transforms, transfer functions, and digital filter design.

5 Phillips, Charles L., and Nagle, H. Troy, *Digital Control System Analysis and Design*, 3rd ed. Englewood Cliffs, N.J.: Prentice-Hall, 1995. This book provides comprehensive coverage of discrete-time control systems analysis and design. It presents many practical examples and makes good use of MATLAB.

6 Shahian, Bahram, and Hassul, Michael, *Control System Design Using MATLAB*. Englewood Cliffs, N.J.: Prentice Hall, 1993. This book provides an introduction to MATLAB programming and uses MATLAB to solve many of the standard problems in classical control and modern control theory.

7 Strum, Robert D., and Kirk, Donald E., *Contemporary Linear Systems Using MATLAB*. Boston: PWS Publishing Co., 1994.

6

Subsystems and Masking

In this chapter, we will explore some of the ways we can use SIMULINK to model more complex systems. We will build hierarchical models and develop custom SIMULINK blocks called masked blocks. We will also discuss conditionally executed subsystems, which can make SIMULINK models much more efficient.

6.1 Introduction

In the preceding chapters, we discussed the basics of building SIMULINK models for continuous, discrete, and hybrid systems. Using the procedures we covered in the preceding chapters, it is possible to model any physical system. However, as your SIMULINK models become more complex, additional SIMULINK capabilities and programming techniques can make the models easier to develop, understand, and maintain. In this chapter, we'll start with a discussion of SIMULINK subsystems, which provide a capability within SIMULINK similar to subprograms in traditional programming languages. Next we'll discuss the use of masking to make subsystems easier to use and understand. Finally, we will discuss conditionally executed subsystems, which facilitate the development of models with multiple modes or phases of operation.

6.2 SIMULINK Subsystems

Most engineering programming languages include the capability to employ *subprograms*. In FORTRAN, there are subroutine subprograms and function subprograms. C subprograms are called functions; MATLAB subprograms are called function M-files. SIMULINK provides an analogous capability called *subsystems*. There are two important reasons for using subprograms: abstraction and software reuse.

As models grow larger and more complex, they can easily become difficult to understand and maintain. Subsystems solve this problem by breaking a large model into a hierarchical set of smaller models. As a simple example, consider the automobile model of Example 4-6. The SIMULINK model is repeated in Figure 6-1. The model consists of two main parts: the automobile dynamics and

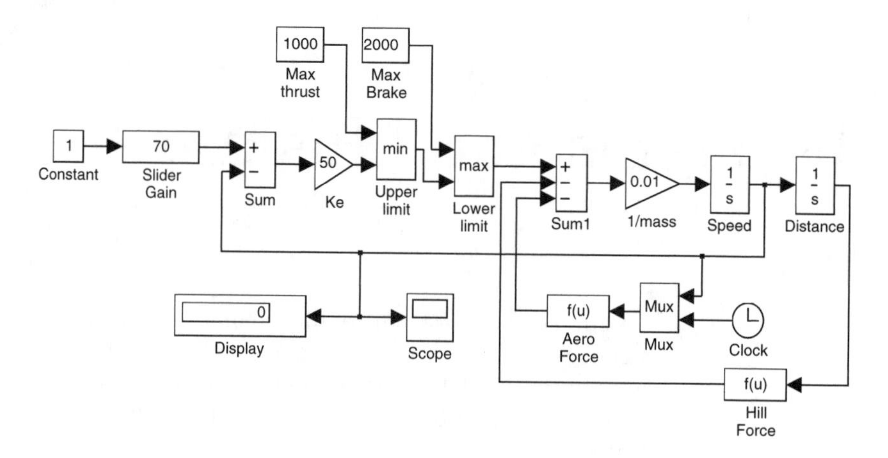

Figure 6-1 Automobile model with proportional speed control

the controller. Examining the model, it is not clear which blocks represent the automobile dynamics and which blocks constitute the controller. In Figure 6-2, we have converted the automobile and controller portions of the model into subsystems. In this version the conceptual structure is clear in the top level of the model (Figure 6-2), but the details of the controller and automobile dynamics are hidden in the subsystems (Figure 6-3). This hierarchical structure is an example of software abstraction.

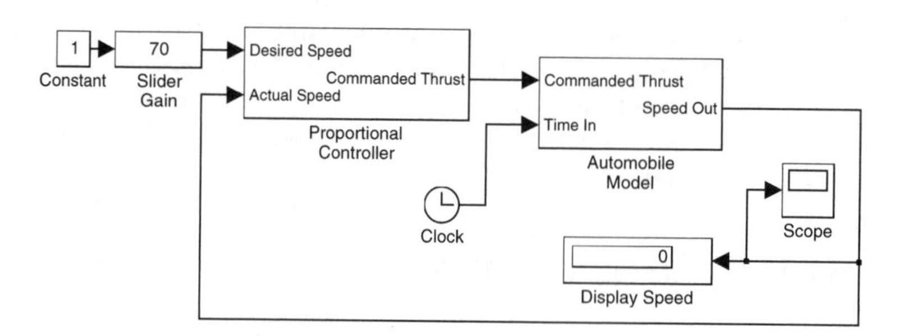

Figure 6-2 Hierarchical automobile model

Subsystems can also be viewed as reusable model components. Suppose that we wish to compare several different controller designs using the same automobile dynamics model. Rather than building a complete new block diagram each time, it is more convenient to build only the part of the model that is new each time—the controller. Not only does this save time building the model, but it also ensures that we are using exactly the same automobile dynamics. An important advantage of software reuse is that once we have verified that a subsystem is correct, we don't have to repeat the testing and debugging process

(a) Controller subsystems

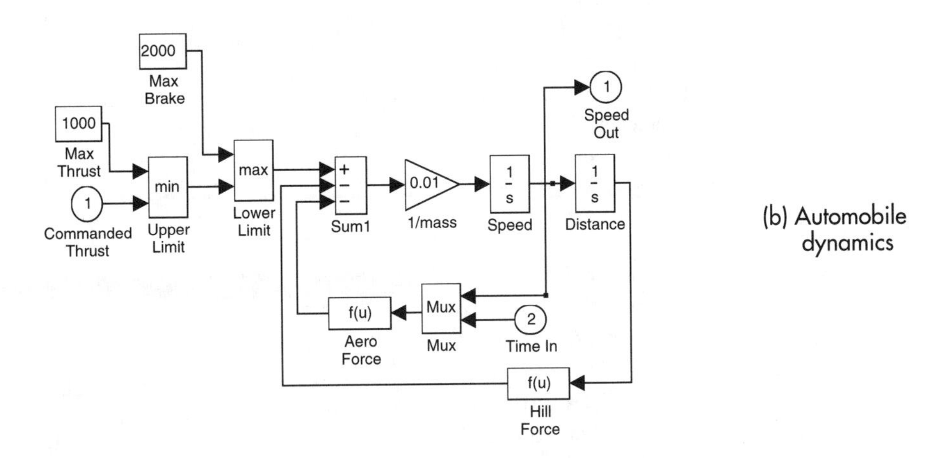

(b) Automobile dynamics

Figure 6-3 Hierarchical automobile model subsystems

each time we use the subsystem in a new model. Subsystems greatly simplify the task of modeling physical systems that contain several instances of a particular component, such as the four tire models that would be required to model the ride characteristics of an automobile.

There are two methods to build SIMULINK subsystems. The first method is to encapsulate a portion of an existing model in a subsystem using **Edit:Create Subsystem**. The second method is to use a Subsystem block from the Connections block library. We'll discuss both methods.

6.2.1 Encapsulating a Subsystem

To encapsulate a portion of an existing SIMULINK model into a subsystem, proceed as follows:

Select all the blocks and signal lines to be included in the subsystem using a bounding box. Note that you must use a bounding box in this instance. It is frequently necessary to rearrange some blocks so that you can enclose only the desired blocks in the bounding box.

Choose **Edit: Create Subsystem** from the model window menu bar. SIMULINK will replace the selected blocks with a Subsystem block that has an input port for each signal entering the new subsystem and an output port for each signal leaving the new subsystem.

SIMULINK will assign default names to the input and output ports.

Resize the Subsystem block so that the port labels are readable, and rearrange the model as desired.

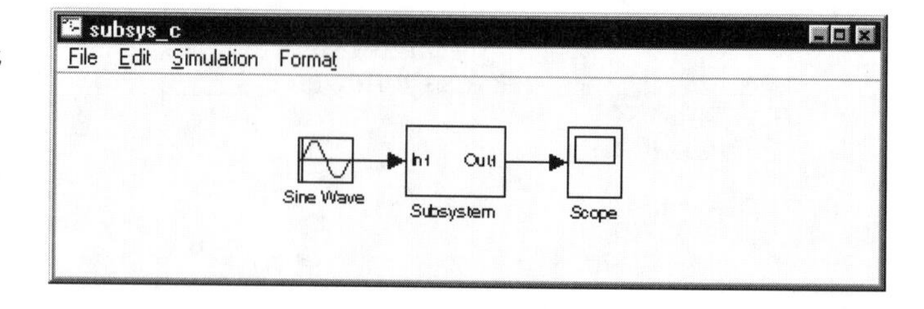

To view or edit the subsystem, double click on the block. A new window will appear, containing the subsystem. In addition to the original blocks, an Inport block is added for the signal entering the subsystem, and an Outport block is added for the signal exiting the subsystem. Changing the labels on these ports changes the labels on the new block's icon. Click on the control to close the subsystem window when you're finished editing the subsystem.

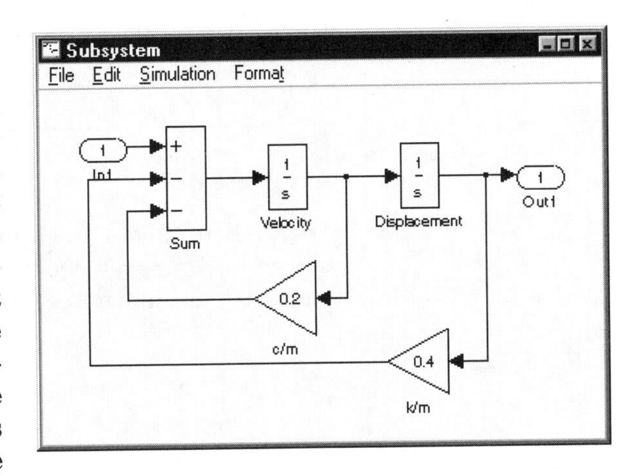

Edit:Create Subsystem does not have an inverse operation. Once you encapsulate a group of blocks into a subsystem, there is no menu choice to reverse the process. Therefore, it is a good idea to save the model before creating the subsystem. If you decide you don't want to accept the newly created subsystem, close the model window without saving and then reopen the model. To reverse manually the encapsulation of a subsystem, copy the subsystem to a new model window, open the subsystem, then copy the blocks from the subsystem window to the original model window.

6.2.2 Subsystem Blocks

If, when building a model, you know that you will need a subsystem, you may find it convenient to build the subsystem in a subsystem window directly. This eliminates the need to rearrange the blocks that will compose the subsystem to fit in a bounding box. It also avoids the need to tidy up the model window after the subsystem is encapsulated.

To create a new subsystem using a Subsystem block, drag a Subsystem block from the Connections block library to the model window. Double click on the Subsystem block. The subsystem window will appear. Build the subsystem using the standard procedures for constructing a model. Use Inport blocks for all signals entering the subsystem, and Outport blocks for all signals leaving the subsystem. If desired, change the labels on the Inport and Outport blocks to identify the purpose of each input and output. Close the subsystem window when you've finished building the subsystem. Note that you do not need to choose **File:Save** before closing the subsystem window; the subsystem is part of the model in which the subsystem is created and is saved when that model is saved.

Example 6-1

We wish to model the spring-mass system composed of carts connected as shown in Figure 6-4. We will build the model from subsystem blocks that model each cart, as shown in Figure 6-5.

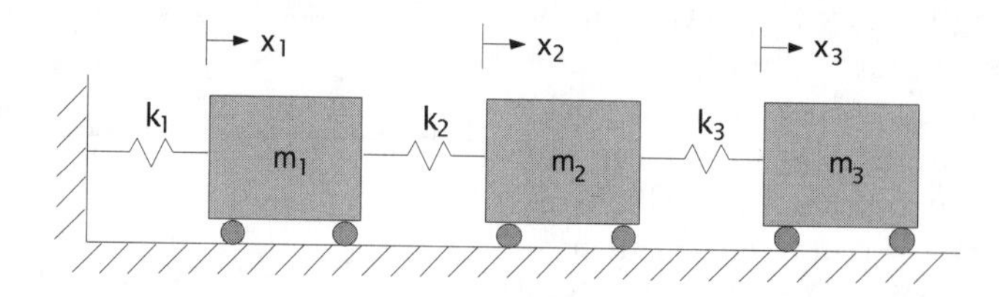

Figure 6-4 Spring-mass system

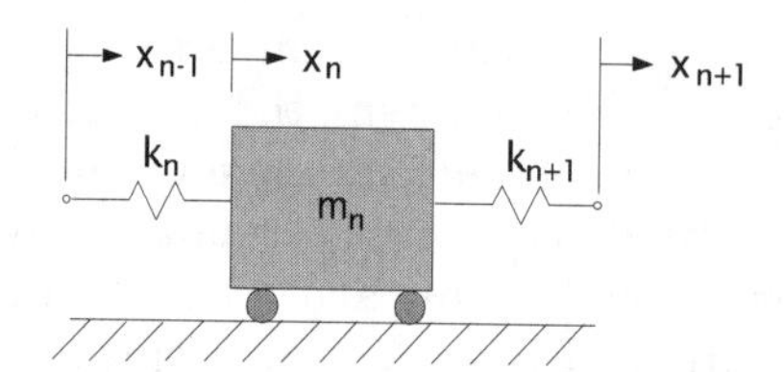

Figure 6-5 Single cart model

The equation of motion for a single cart is

$$\ddot{x}_n = \frac{1}{m_n}[k_n(x_{n-1} - x_n) + k_{n+1}(x_n - x_{n+1})]$$

Using the procedure discussed in Section 6.2.2, construct the subsystem as shown in Figure 6-6. This subsystem will model cart 1. The inputs to the single cart subsystem are x_{n-1} (position of the cart to the left) and x_{n+1} (position of the cart to the right). The subsystem output is x_n (position of cart). Notice that each spring is referenced in two Subsystem blocks, one for the cart to the right of the spring and one for the cart to the left.

Once the subsystem is complete, close the subsystem window. Make two copies of the Subsystem block, and connect the blocks as shown in Figure 6-7.

It is convenient in this case to enter the spring constants (`k1, k2, k3`) and cart masses (`m1, m2, m3`) as MATLAB variables, and assign values to the variables

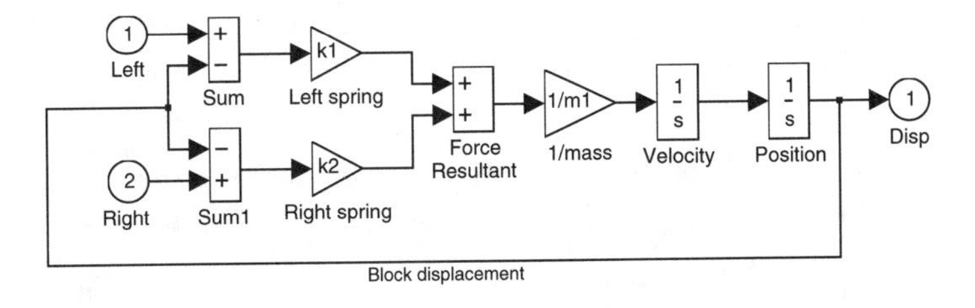

Figure 6-6 Cart model subsystem for cart 1

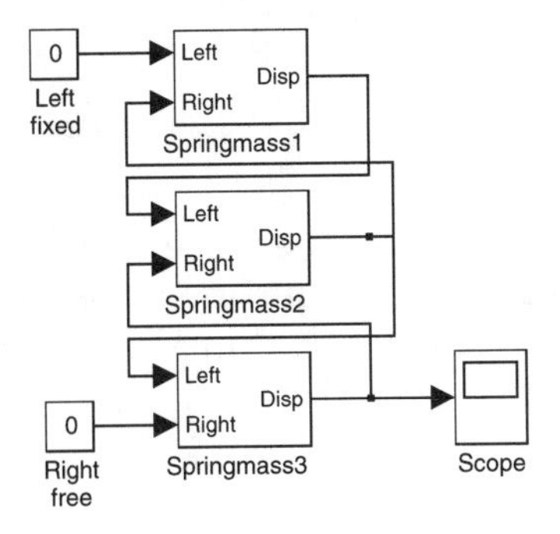

Figure 6-7 Three cart model using subsystems

using a MATLAB script M-file (which we named `set_x4a.m`), as shown in Figure 6-8. Execute this script M-file from the MATLAB prompt before running the simulation. (Note that the script M-file (extension `.m`) must have a name that is different from the name of the SIMULINK model. For example, if the SIMULINK model is named `examp_1.mdl` and you name the script M-file `examp_1.m`, MATLAB will open the SIMULINK model when you enter the command `examp_1` at the MATLAB prompt.)

The block parameters for each block in each copy of the subsystem must now be set. For cart 1, set the value of **Gain** for the Gain block labeled `Left Spring` to `k1`, and for Gain block Right Spring to `k2`. Next set **Gain** for the Gain block labeled 1/mass to `1/m1`. Initialize the Velocity Integrator block to `0`, and the Position Integrator block to `1`.

For cart 2, set the value of **Gain** for the Gain block labeled Left Spring to `k2`, and for Gain block Right Spring to `k3`. Next set **Gain** for the Gain block labeled 1/mass to `1/m2`. Initialize the Velocity Integrator block to `0`, and the Position Integrator block to `0`.

For cart 3, set the value of **Gain** for the Gain block labeled Left Spring to `k3`, and for Gain block Right Spring to `0` since there is no right spring for this cart.

Next set **Gain** for the Gain block labeled 1/mass to `1/m3`. Initialize the Velocity Integrator block to `0`, and the Position Integrator block to `0`.

```
% Set the spring constants and block mass values
k1 = 1 ;
k2 = 2 ;
k3 = 4 ;
m1 = 1 ;
m2 = 3 ;
m3 = 2 ;
```

Figure 6-8 MATLAB script `set_x4a.m` to initialize spring constants

We configured the Scope block to save the scope data to the workspace, and set the simulation **Start time** to `0` and **Stop time** to `100`. After running the simulation, the scope data was plotted from within MATLAB, resulting in the trajectory plot for cart 3 shown in Figure 6-9.

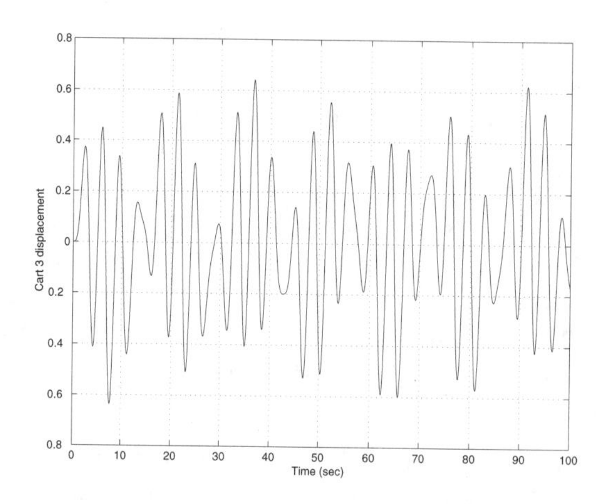

Figure 6-9 Trajectory of cart 3

6.3 Masked Blocks

Masking is a SIMULINK capability that extends the concept of abstraction. Masking permits us to treat a subsystem as if it were a simple block. A masked block may have a custom icon, and it may also have a dialog box in which configuration parameters are entered in the same way parameters are entered for blocks in the SIMULINK block libraries. The configuration parameters may be used directly to initialize the blocks in the underlying subsystem, or they may be used to compute data to initialize the blocks.

To understand the concept of masking, consider the model shown in Figure 6-10a. This model is equivalent to the model in Example 6-1, but it is easier to use. Double clicking on the block labeled `Spring-mass 1` opens the dialog box shown in Figure 6-10b. Instead of opening the dialog box for each Gain block and each Integrator to set the block parameters, you can enter all the parameters for each subsystem in the subsystem's dialog box. The dialog box in Figure 6-10b "masks" a subsystem that is nearly identical to the subsystem in Figure 6-6.

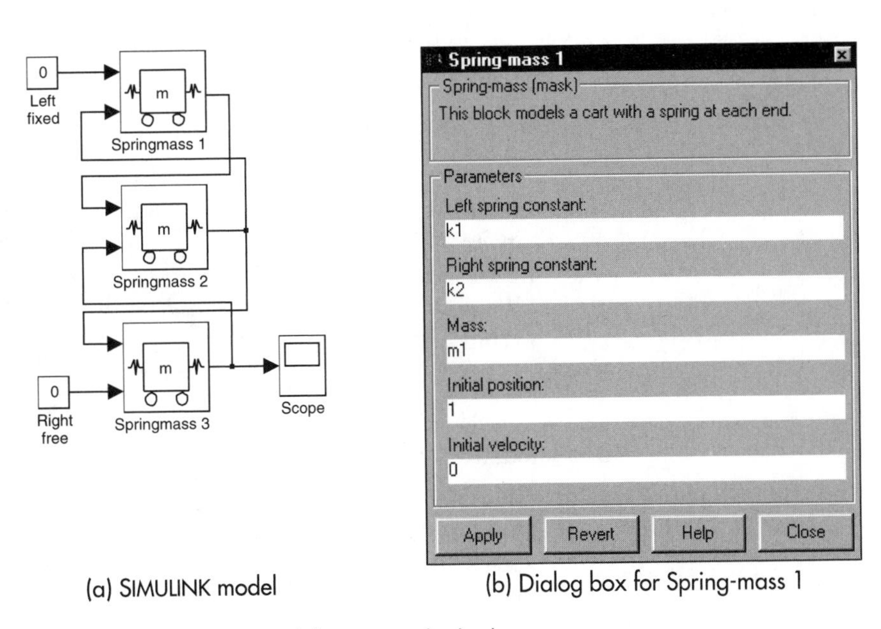

(a) SIMULINK model (b) Dialog box for Spring-mass 1

Figure 6-10 Three-cart model using masked subsystems

In this section, we will explain the steps in creating a masked subsystem. The examples will show how to create the spring-mass masked subsystem. Additional examples will illustrate other masking features.

The process of producing a masked block can be summarized as follows:

1 Build a subsystem using the procedures discussed in Section 6.2.

2 Select the subsystem block and then choose **Edit:Create Mask** from the model window menu bar.

3 Using the Mask Editor, set up the mask documentation, dialog box, and, optionally, build a custom icon.

6.3.1 Converting a Subsystem into a Masked Subsystem

The first step in creating a masked subsystem is to create a subsystem using the procedures described in Section 6.2. To illustrate the process, let's build a Spring-mass masked block starting with one of the subsystems in the model in Figure 6-7.

Open the model shown in Figure 6-7. Next open a new model window. Drag a copy of the block labeled `Spring-mass 1` to the new model window.

Select the block and then choose **Edit:Create Mask** from the new model window menu bar.

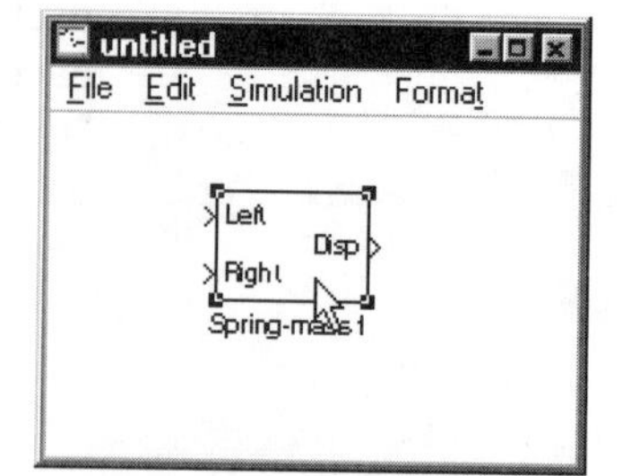

The Mask Editor dialog box will appear. Note that the Mask Editor has three tabbed pages. We will discuss each page in the following subsections.

Before proceeding, save the new model window using the name `spm_msk`.

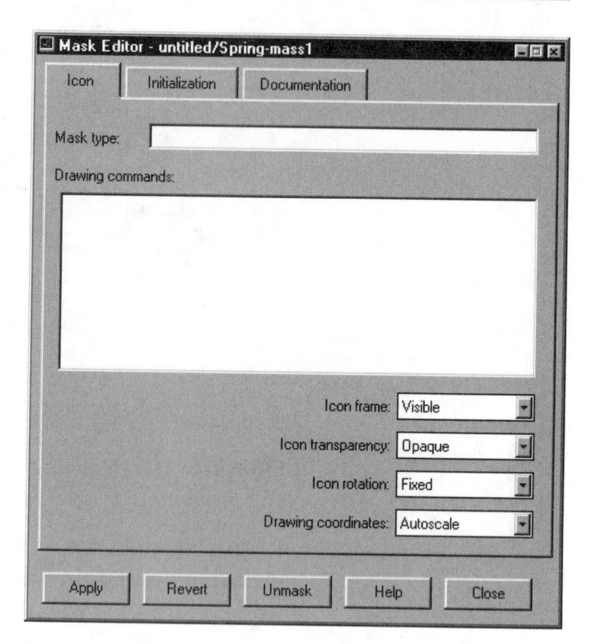

6.3.2 Mask Editor Documentation Page

The Documentation page is illustrated in Figure 6-11a. The page consists of three fields, which in Figure 6-11a have been filled in for the Spring-mass block. All fields in the Documentation page are optional.

After filling in the fields as shown, select **Close**. Then double click on the Spring-mass block, opening the dialog box shown in Figure 6-11b. You can

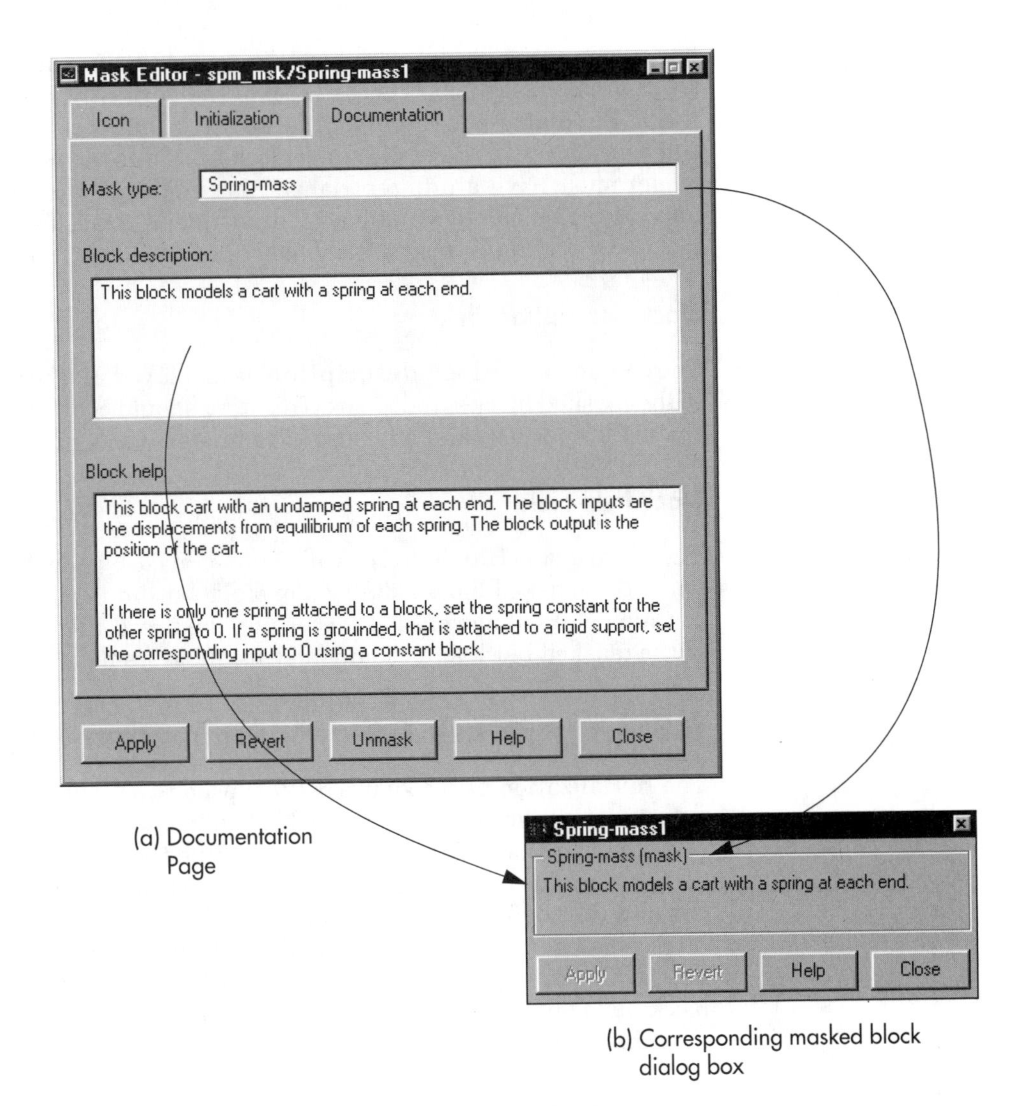

Figure 6-11 Mask Editor Documentation page

return to the Mask Editor by selecting the block and then choosing **Edit:Edit Mask** from the model window menu bar.

We will discuss each field in the Documentation page next.

Mask type Field

The contents of the first field, **Mask type**, will be displayed as the block type in the masked block's dialog box. Notice that there are two labels in the upper left corner of the block dialog box (Figure 6-11b). The label in the window title bar (here `Spring-mass1`) is the label of the currently selected block. The label inside the dialog box (here `Spring-mass`) is the block type. Every instance of this new block will have the same block type, but each instance in a particular model must have a different label. Also, note the word `mask` in parentheses appended to block type, indicating that this is a masked block. (Compare the masked block dialog box to the dialog box for a block from a block library.)

Block description Field

The second field, **Block description**, is displayed in a bordered area at the top of the masked block's dialog box. This area should contain a brief description of the block's purpose and any needed reminders concerning the use of the block.

Block help Field

The third field, **Block help**, will be displayed by the MATLAB help system when the masked block's dialog box **Help** button is pressed. This field should contain detailed information concerning the use, configuration, and limitations of the masked block and the underlying subsystem.

6.3.3 Mask Editor Initialization Page

The Initialization page (Figure 6-12) is used to set parameters of blocks in the subsystem underlying the masked block. The Initialization page can be divided into three sections. The top section contains the **Mask type** field. The center section contains a set of fields that define the fields in the masked block's dialog box and that define a local variable corresponding to each field in the masked block's dialog box. The bottom section contains the **Initialization commands** field, which can be used to define additional variables to set parameters in the block dialog box or to be used on the Icon page (to be discussed later). We will describe each section.

Mask type Field

The top section of the page contains the **Mask type** field, which is identical to the **Mask type** field on the Documentation page. **Mask type** may be entered or edited on this page, or on the other Mask Editor pages. Changing **Mask type** on one page changes it on the other pages as well.

Block Dialog Box Prompt Section

The center section of the Mask Editor Initialization page is used to create, edit, and delete dialog box fields. This section consists of a scrolling list of dialog box fields, buttons to add, delete, and move fields, and five fields used to configure

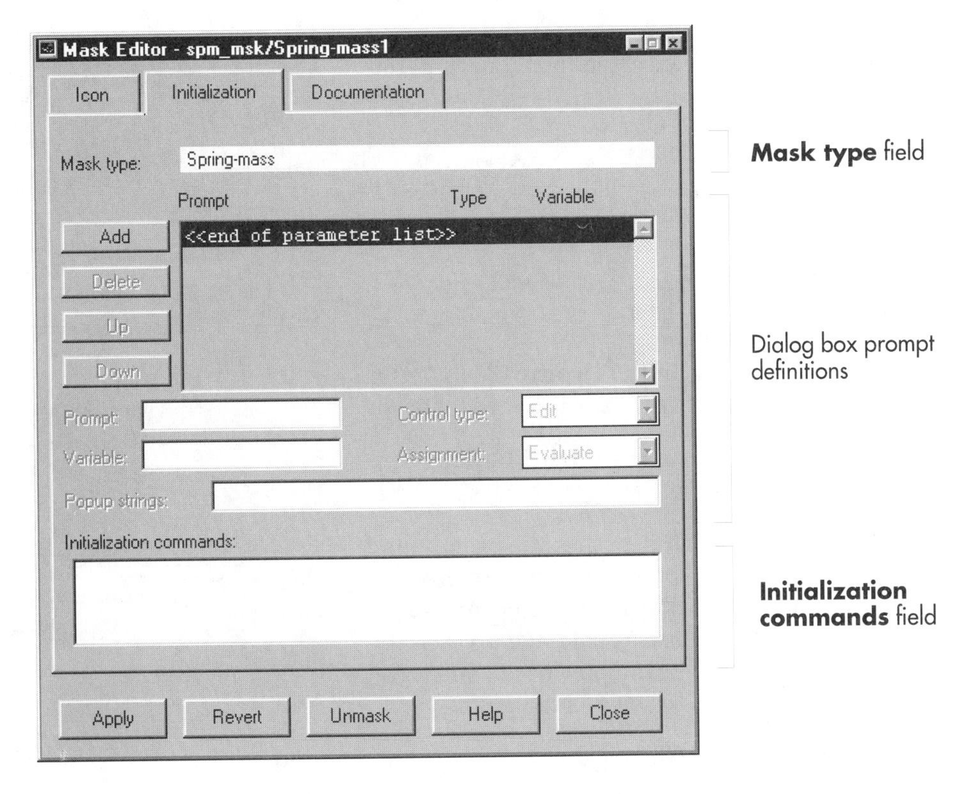

Figure 6-12 Blank Initialization page

the block dialog box. To add the first prompt field in the Spring-mass block dialog box, proceed as follows:

Select the block and then open the Mask Editor by choosing **Edit:Edit Mask** from the model window menu bar. Select the Initialization page.

Click on `<<end of parameter list>>` and then click on **Add**. (Note that here only the Prompt section of the Initialization page is shown.)

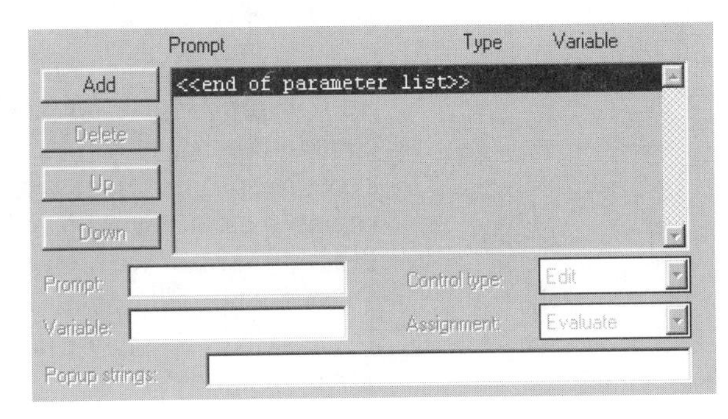

A blank line will be inserted in the parameter list.

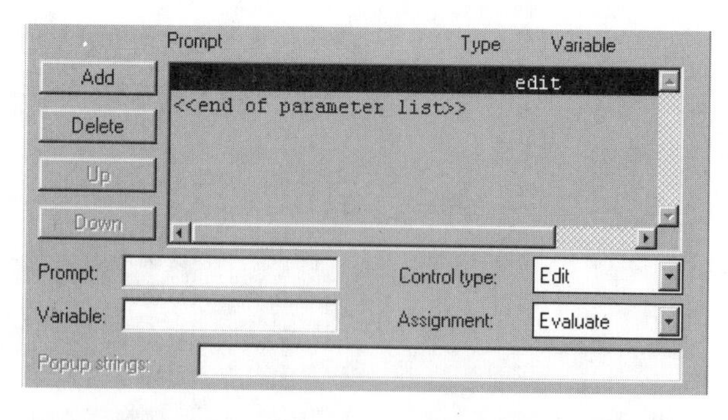

In the **Prompt** field, enter `Left spring constant`.

In the **Variable field**, enter `k_left`.

Notice that the prompt and variable name are displayed in the parameter list. You'll have to scroll the parameter list to the right to see all of `k_left`.

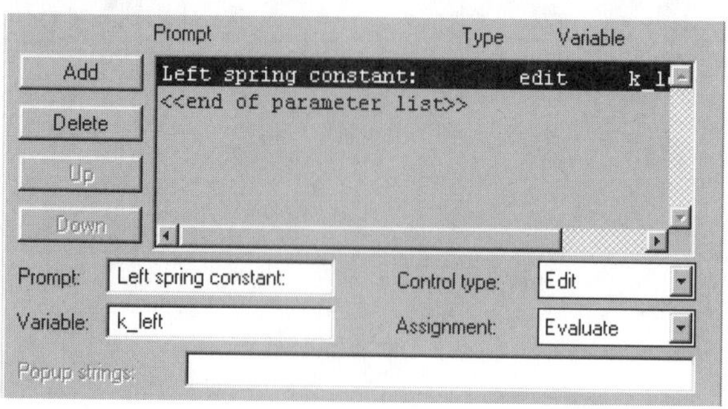

Choose **Close**.

Double click on the Spring-mass subsystem. The block dialog box now has a prompt. The value entered in the field corresponding to the prompt will be assigned to MATLAB variable `k_left`.

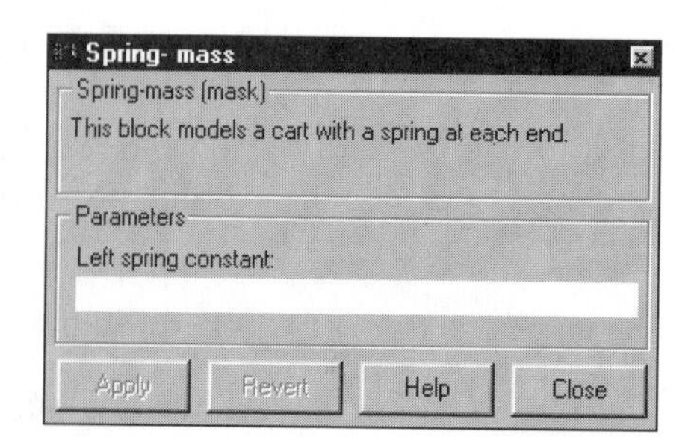

Add prompt `Right spring constant` with variable `k_right`, prompt `Mass` with variable `mass`, prompt `Initial position` with variable `x0`, and prompt `Initial velocity` with variable `x_dot0`.

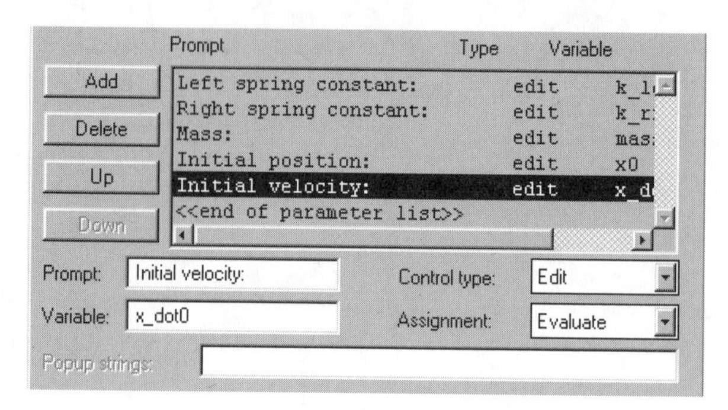

That completes the process of creating the dialog box fields. To see the results, press **Close** to save the changes and exit the Mask Editor.

Double click on the Spring-mass block, opening the block dialog box, which should look like this.

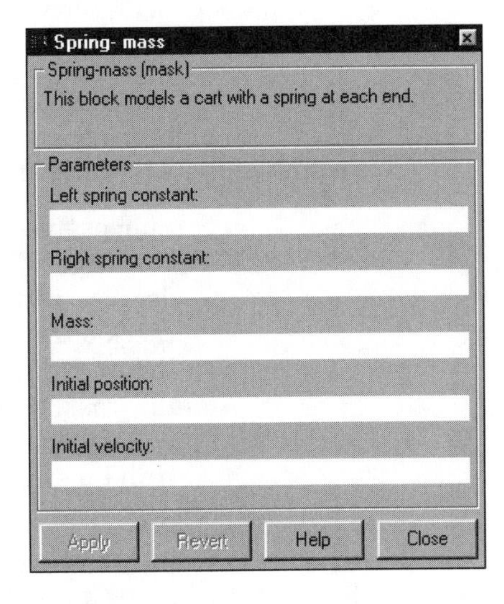

Now we will discuss in more detail the buttons and fields in the dialog box prompt section of the Mask Editor. First, notice that there are four buttons on the left side of the dialog box prompt section. These buttons are used to add, delete, and arrange dialog box prompts. To create a new field, click on the line in the scrolling parameter list of the item you wish to follow the new field. If you wish for the new field to be the bottom field, click on `<<end of parameter list>>`. Then press **Add**. A blank line will appear in the scrolling list. Pressing the **Delete** button deletes the selected field. **Up** and **Down** move the selected field in the appropriate direction in the list of fields.

Field **Control type** is a drop-down list with three options: **Edit**, **Checkbox**, and **Popup**. **Edit** produces a field in which data is entered (a fill-in-the-blank field) and is the most common type of field. **Checkbox** generates a field that has two possible values, depending on whether the box is checked or not. **Popup** produces a list of choices set in **Popup strings**.

The value assigned to the internal variable associated with a block dialog box field will depend on the contents of the **Assignment** field. **Assignment** may be set to **Evaluate** or **Literal**. If **Evaluate** is chosen, the variable associated with the field will contain the value of the expression in the field. So, for example, if the field contains `k1`, and in the MATLAB workspace `k1` is assigned the value `2.0`, the variable associated with the dialog field will be assigned the value `2.0`. If **Literal** is chosen, the variable associated with the dialog field will contain the character string `'k1'`.

Setting **Control type** to **Checkbox** produces a check box field. The variable associated with a check box field will be assigned a value depending on the setting of **Assignment**. If **Assignment** is **Evaluate**, the variable associated with the field will be set to `0` if not checked, or `1` if checked. If **Assignment** is **Literal**, the variable associated with the field will be set to `'no'` if not checked, and `'yes'` if checked.

Example 6-2

Suppose we wish to configure a masked subsystem such that its block dialog box has a check box allowing the user to specify that angular inputs are to be in degrees rather than radians. The value of the variable associated with the check box (`c_stat`) is to be `0` if the box is not checked, and `1` if the box is checked. Figure 6-13a shows the Mask Editor prompt section configured to accomplish this task, and Figure 6-13b shows the corresponding masked block dialog box. In this example, set **Mask type** to `Check box example`, and **Block description** (on the Documentation page) to `This block illustrates a check box.`

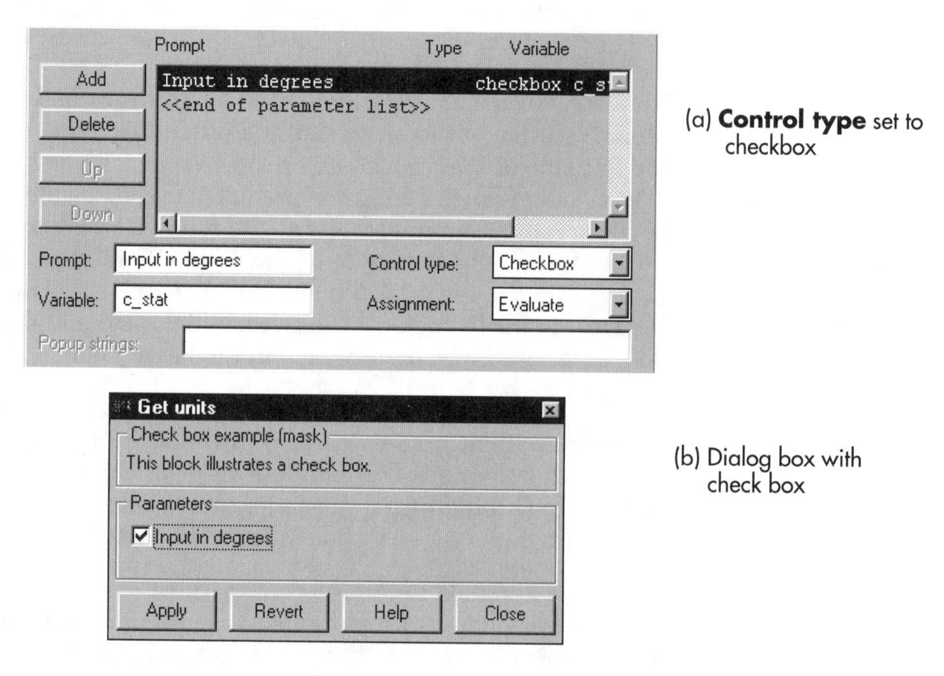

(a) **Control type** set to checkbox

(b) Dialog box with check box

Figure 6-13 Check box field

If **Control type** is set to **Popup**, the field **Popup strings** is used to define a list of choices, as will be shown in Example 6-3. The variable associated with a popup field will be assigned a value depending on the setting of **Assignment**.

If **Assignment** is set to **Evaluate**, the variable associated with the field will be set to the ordinal number of the selected popup choice. So for example, if the first choice is selected, the value of the variable will be set to 1. If the second choice is selected, the value of the variable will be set to 2, and so on. If **Assignment** is set to **Literal**, the variable will contain the character string corresponding to the selected choice. The list of choices for a popup field is set in **Popup strings**. The choices are entered in sequence, separated with the pipe symbol (|). For example, if the choices are to be `Very hot, Hot, Warm, Cool,` and `Cold`, **Popup strings** should contain `Very hot|Hot|Warm|Cool|Cold.`

Example 6-3

Suppose we wish to create a masked block that produces an output signal defined by a popup list containing choices `Very hot`, `Hot`, `Warm`, `Cool`, and `Cold`. To do this, open a new model and then drag a Constant block into the model window. Select the Constant block using a bounding box and then choose **Edit:Create Subsystem** from the model window menu bar. Select the new subsystem and then choose **Edit:Create Mask** from the model window menu bar. Set **Mask type** to `Popup example` and **Block description** to `This block illustrates a popup list.` Configure the prompt section of the Mask Editor Initialization page as shown in Figure 6-14a. Choose **Close**. Double click on the masked block and then click on the popup control. The block dialog box should be as shown in Figure 6-14b. In Example 6-6 we will show how to configure the **Initialization commands** field for this block.

Initialization commands

The bottom section of the Initialization page is the field **Initialization commands**. This field can contain one or more MATLAB statements that assign values to MATLAB variables used to configure blocks in the masked subsystem. The MATLAB statements may use any of MATLAB's operators, built-in or user-written functions, and control flow statements such as `if, while,` and `end`. The scope of variables in the **Initialization commands** field is local; variables defined in the MATLAB workspace are not accessible.

The **Initialization commands** field displays only four lines but may contain many lines. You can move up and down in the field using the cursor keys.

Each command in the **Initialization commands** field should normally be terminated with a semicolon (;). If you omit the semicolon for a command, the results of the command will be displayed in the MATLAB window whenever the command is executed. This provides a convenient means to debug the commands.

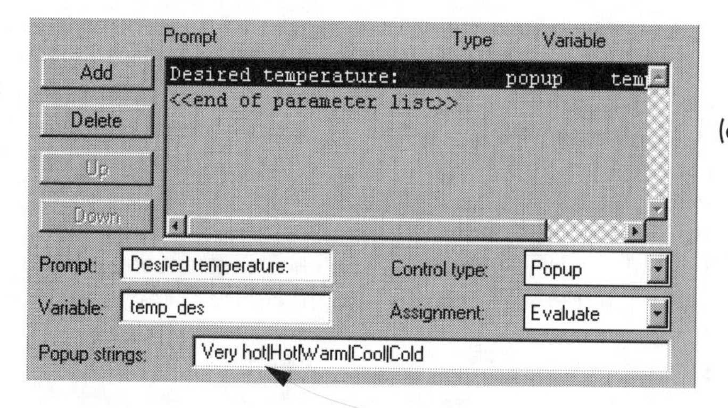

(a) Prompt section of Mask Editor Initialization page

Popup strings defines the popup menu choices

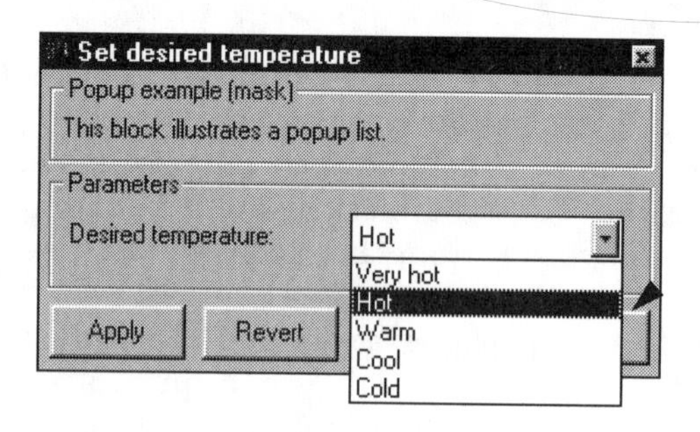

(b) Block dialog box with popup menu

Figure 6-14 Popup menu

Example 6-4

The Spring-mass block icon will need two wheels. To prepare to draw the wheels, create two vectors, one containing the x-coordinates of a small circle and the other the y-coordinates. These variables will be used in the Icon page to draw the wheels.

Configuring Subsystem Blocks

The blocks in the masked subsystem must be configured to use the variables defined on the Initialization page. To configure the blocks in the Spring-mass subsystem, select the subsystem and then choose **Edit:Look Under Mask** from the model window menu bar. Double click on the Gain block labeled `Left spring`, and set **Gain** to `k_left`. Likewise, set the value of **Gain** for Gain block `Right spring` to `k_right` and for Gain block `1/mass` to `1/mass`. Set **Initial condition** for Integrator `Velocity` to `x_dot0`, and Integrator `Position` to `x0`.

The subsystem should now appear as shown in Figure 6-15. Close the subsystem and save the model.

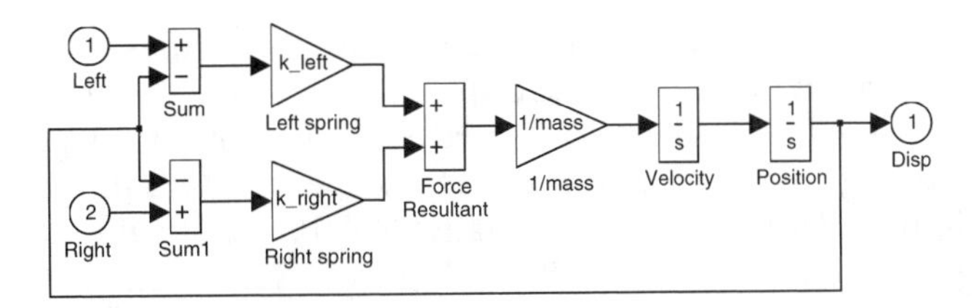

Figure 6-15 Spring-mass subsystem configured to use masked block variables

Local Variables

An important difference between masked subsystems and nonmasked subsystems is the scope of variables in the dialog boxes for the blocks in a subsystem. Blocks in nonmasked subsystems may use any MATLAB variable currently defined in the MATLAB workspace. This feature was used to initialize the subsystems in Example 6-1. The blocks in a masked subsystem can't access variables in the MATLAB workspace; a masked subsystem has its own internal name space that is independent of the MATLAB workspace and all other masked subsystems in a SIMULINK model. This is an extremely valuable feature of masked subsystems because it eliminates the possibility of unintentional variable name conflicts.

A masked subsystem's internal variables are created and assigned values using the Mask Editor dialog fields and initialization commands. Each dialog field in a masked block's dialog box defines an internal variable accessible only within the masked subsystem. Additional internal variables may be defined in the **Initialization commands** field of the Initialization page.

The connections between the MATLAB workspace and a masked subsystem are the contents of the masked block's dialog box fields. An input field in a masked block's dialog box may contain constants or expressions using variables defined in the MATLAB workspace. The value of the contents of the input field is assigned to the masked subsystem internal variable associated with the input field. This internal variable may be used to initialize a block in the masked subsystem, or it may be used to define another internal variable defined in the **Initialization commands** field.

Consider the model shown in Figure 6-10, with the masked subsystem configured as shown in Figure 6-15. The spring constant for the left spring in each instance of the masked subsystem is `k_left`. However, the contents of `k_left` in each instance is different. `k_left` for the first Spring-mass block should be set (using the Spring-mass block dialog box) to `k1`. `k_left` for the second should be set to `k2`, and for the third Spring-mass block to `k3`.

Example 6-5

To illustrate the use of internal variables in a masked subsystem, consider the assignment of a value to the spring constant of the right spring in the cart subsystem of Figure 6-10. In Figure 6-12 the contents of the cart subsystem dialog field **Right spring constant** are associated with internal variable `k_right`. In Figure 6-16 we see that the Gain block labeled `Right spring` is set to `k_right`. Thus when the M-file script in Figure 6-8 is executed, the value of the gain for the Gain block `Right spring` in this particular instance of the cart subsystem is set to 2.

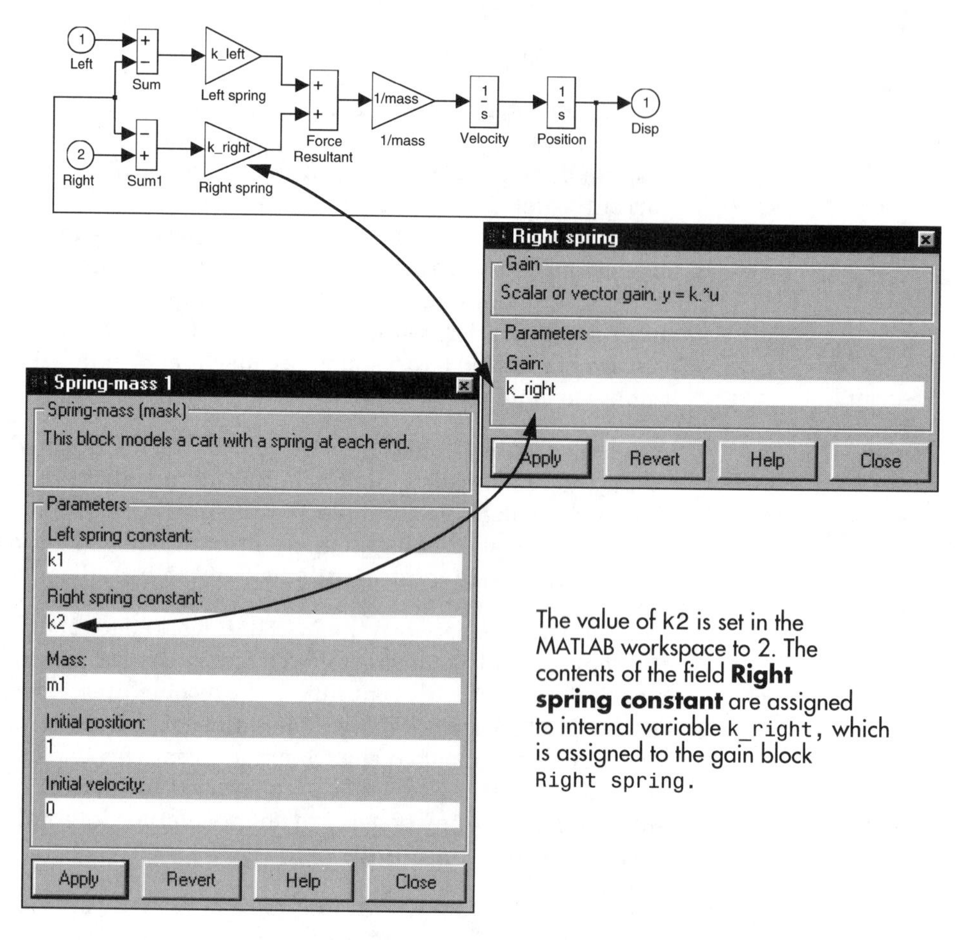

Figure 6-16 Right spring constant gain block initialization

Example 6-6

For the masked subsystem in Example 6-3, set a variable `temp_val` as follows:

Menu choice	`temp_val`
Very hot	120
Hot	100
Warm	85
Cool	70
Cold	50

To accomplish this task, place the following statements in the **Initialization commands** field:

```
temp_list = [120,100,85,70,50] ;
temp_val = temp_list(temp_des) ;
```

The first statement creates a vector of temperatures. The second statement uses `temp_des` (the variable associated with the popup list in Example 6-3; Figure 6-14) as an index into the vector. Choose **Close** and then, with the masked block still selected, choose **Edit:Look Under Mask** and set the Constant block dialog box field **Constant value** to `temp_val`.

6.3.4 Mask Editor Icon Page

The Icon page allows you to design custom icons for masked blocks. The Icon page used to create the custom icon for the cart block in Figure 6-10 is shown in Figure 6-17. (Recall that `x` and `y` were defined in the **Initialization commands** field in Example 6-4.) The page consists of six fields. The top field, **Mask type**, is identical to the **Mask type** field on the other two Mask Editor pages. **Drawing commands** is a multiple-line field in which we enter one or more MATLAB statements to draw and label the icon. The remaining four fields configure the block icon. It will be easier to explain **Drawing commands** if we first discuss the configuration fields.

Icon frame Field

The first icon configuration field, **Icon frame**, is a drop-down list containing two choices: **Visible** and **Invisible**. The icon frame is the border of the block icon. Figure 6-18 illustrates the Spring-mass block with and without the icon frame.

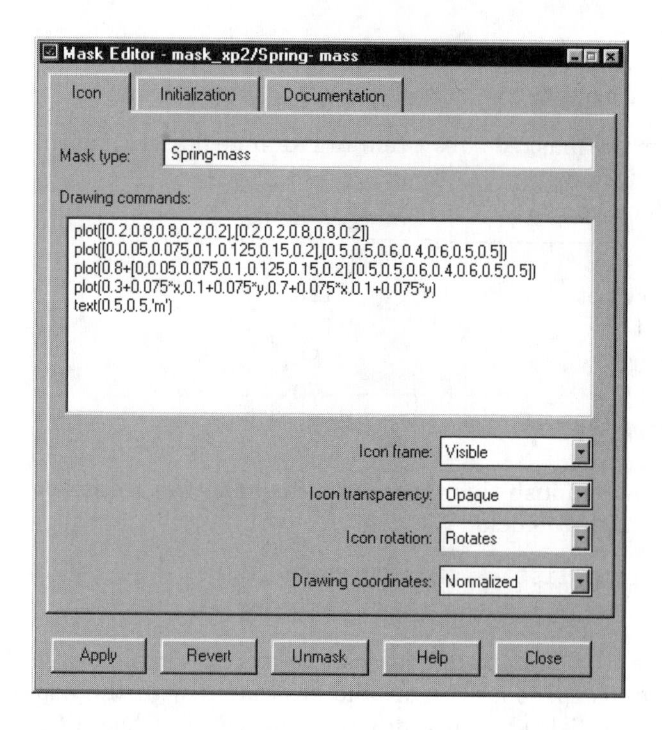

Figure 6-17 Mask Editor Icon page for the cart subsystem

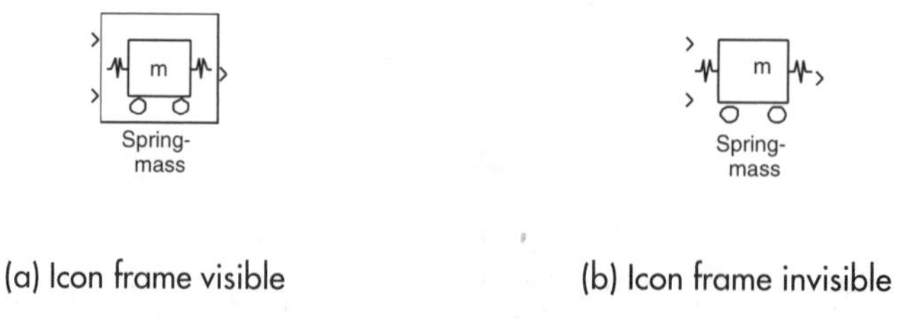

(a) Icon frame visible (b) Icon frame invisible

Figure 6-18 Icon frame visibility

Icon transparency Field

The second icon configuration field, **Icon transparency**, is a drop-down list containing two choices: **Transparent** and **Opaque**. Figure 6-19 shows the Spring-mass block with **Icon transparency** set to both options. Note that when **Transparent** is selected, the labels on the Inport and Outport blocks in the subsystem underlying the mask are visible. Selecting **Opaque** hides the labels.

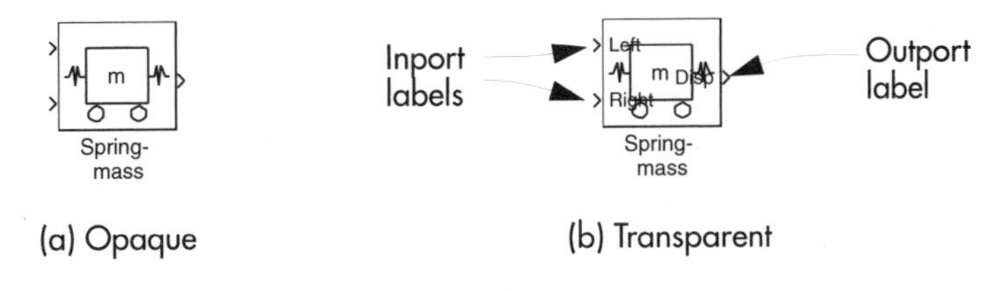

(a) Opaque (b) Transparent

Figure 6-19 Icon transparency

Icon rotation Field

The third icon configuration field, **Icon rotation**, is a drop-down list containing two choices: **Fixed** and **Rotates**. This field determines the behavior of the block icon when **Format:Flip block** and **Format:Rotate block** are selected. If **Fixed** is selected, when the block is rotated or flipped, the icon orientation doesn't change. When **Rotates** is selected, the orientation of the icon is the same as the orientation of the block. Figure 6-20 illustrates the difference. **Fixed** is frequently desirable, particularly in cases in which the block icon contains text.

(a) **Rotates** selected (b) **Fixed** selected

Figure 6-20 Icon rotation

Drawing coordinates Field

The final icon configuration field, **Drawing coordinates**, determines the scale used in plotting icon graphics and locating text on the icon. The field is a drop-down list consisting of three choices: **Pixel**, **Autoscale**, and **Normalized.**

Pixel is an absolute scale and will result in an icon that isn't resized when the block is resized. The coordinates of the lower left corner of the icon are (0,0). The units are pixels, and therefore the size of the icon will depend on the display resolution. Figure 6-21 shows a masked block with **Drawing coordinates** set to **Pixel.**

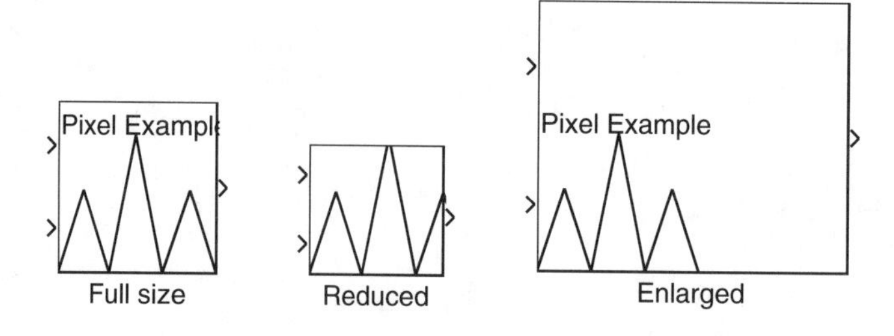

Figure 6-21 Drawing coordinates set to Pixel

Autoscale adjusts the size of the icon to fit exactly in the block's frame (even if the frame is invisible). Figure 6-22 shows the masked block in Figure 6-21 reset to **Autoscale**. Note that the text on the icon does not change size when the block is resized.

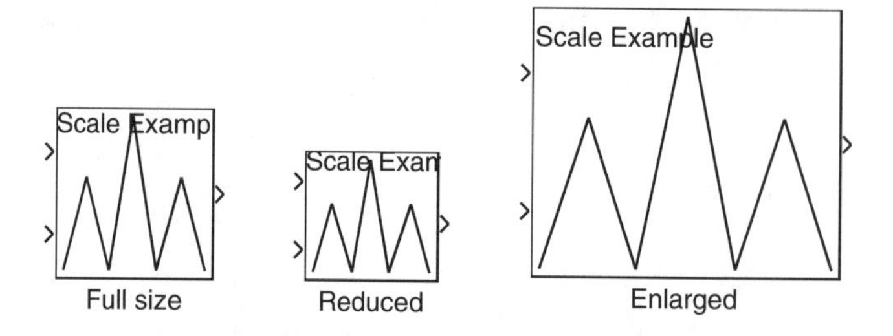

Figure 6-22 Drawing coordinates set to Autoscale

Normalized specifies that the drawing scale is 0.0 to 1.0 in both the horizontal and vertical axes. The coordinates of the lower left corner of the icon (in its default orientation, not rotated or flipped) are defined to be (0,0), and the coordinates of the upper right corner of the icon are (1,1). When the block is resized, the coordinates are also resized. Text does not change size when the block is resized. Figure 6-23 illustrates a masked block with **Drawing coordinates** set to **Normalized**.

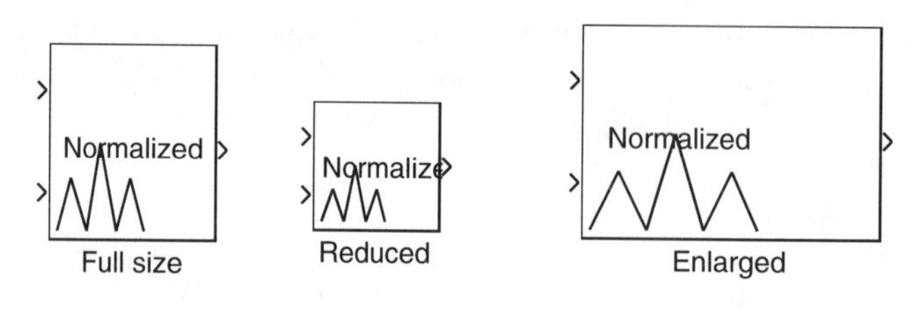

Figure 6-23 Drawing coordinates set to Normalized

Drawing commands

Several different MATLAB statements may be entered in **Drawing commands** to customize a block's icon. Table 6-1 lists these commands.

Table 6-1 Icon drawing commands

Command	Description
`disp(string)`	Display *string* in the center of the icon.
`text(x,y,string)`	Display `string` starting at (*x*,*y*).
`fprintf(string,list)`	Display the results of the `fprintf` statement at the center of the icon.
`plot(x_vector,y_vector)`	Draw a plot on the icon.
`dpoly(num,denom)`	Display a transfer function centered on the icon.
`dpoly(num,denom,'z')`	Display a discrete transfer function in ascending powers of z.
`dpoly(num,denom,'z-')`	Display a discrete transfer function in descending powers of z.
`droots(zeros,poles,gain)`	Display a transfer function in zero, pole, gain format.

Three of the commands display text on the icon. The simplest, `disp(string)`, displays `string` centered on the icon. This command is useful for placing a simple descriptive title in the center of the icon. The `text(x,y,string)` command permits you to locate a string anywhere on the icon, using the icon coordinate system as specified in the **Drawing coordinates** field. The third text command, `fprintf(string,list)`, is identical to the `fprintf` statement. (Enter `help fprintf` at the MATLAB prompt for details on `fprintf`.) Using `fprintf`, you can build labels that use variables defined by the block's dialog fields and the statements in the field **Initialization commands** on the Initialization page. Embed a newline character (`\n`) in the string to produce a label with multiple lines. Like the `disp` command, `fprintf` places the label at the center of the icon.

A character string used to display text on the icon may be a literal string, or it may be a MATLAB string variable. A literal string is a sequence of printable characters enclosed in single quotes. For example, to place the label "Special Block" in the center of a block's icon, use the command

```
disp('Special Block')
```

A string variable is a MATLAB variable that represents a character string instead of a number. MATLAB provides several functions for building and manipulating string variables. These functions can be used in the field **Initialization commands** on the Initialization page to create strings for use in the text display commands. A particularly useful string function is `sprintf`. `sprintf` is similar to `fprintf`, but it writes to a character string instead of the screen or a file. An excellent reference for a detailed discussion of MATLAB string variables and functions is the text by Hanselman and Littlefield [1].

Example 6-7

Suppose we wish to change the icon shown in Figure 6-10 such that the cart mass is displayed on the icon just above the wheels, and the units of mass are kilograms. Add the following command to the **Initialization commands** field on the Initialization page of the Mask Editor:

```
b_label =sprintf('%1.1f kg',m);
```

Then enter the following command in the **Drawing commands** field of the Icon page:

```
text(0.3,0.35,b_label);
```

The block icon will be as shown in Figure 6-24.

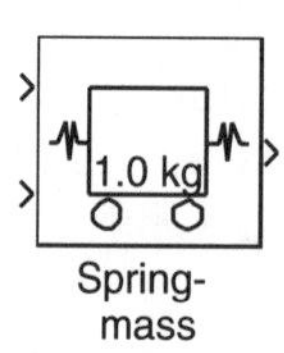

Figure 6-24 Displaying the cart mass on its icon

The `plot(x_vector,y_vector)` command displays graphics on the block icon. This command is similar to the MATLAB `plot` command, but has fewer options. Mask Editor `plot` command does not support options to set line styles or colors, and will not plot two-dimensional arrays. The command expects pairs of vectors specifying sequences of x- and y-coordinates. There may be more than one pair of vectors in a single `plot` command, and there may be more than one `plot` command for an icon. Figure 6-17 shows the commands used to create the cart icon.

Example 6-8

Suppose we wish to display sine and cosine functions on an icon. Placing the following commands in the **Initialization commands** field will produce the necessary vectors:

```
x_vector = [0:0.05:1] ;
y_sin = 0.5 + 0.5*sin(2*pi*x_vector) ;
y_cos = 0.5 + 0.5*cos(2*pi*x_vector) ;
```

Next place the following command in **Drawing commands**:

```
plot(x_vector,y_sin,x_vector,y_cos)
```

The block will appear as shown in Figure 6-25.

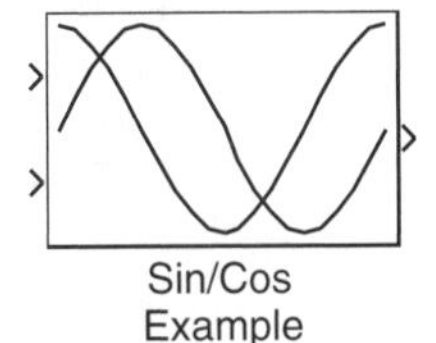

Figure 6-25 Drawing curves on a block icon

Example 6-9

The SIMULINK Logical Operator block can be configured to implement AND and OR gates, but the block icon is a rectangle with the logical operator it implements displayed on the icon. By placing copies of the Logical Operator block into masked subsystems, we can produce AND and OR blocks that more closely resemble the conventional icons for these gates.

To produce the AND gate block, start with a Logical Operator block configured to implement AND. Select the block using a bounding box and then choose **Edit:Create subsystem** from the model window menu bar. Select the subsystem and then choose **Edit:Create mask** from the model window menu bar. Open the Mask Editor and place the following command in the **Initialization commands** field on the Initialization page:

```
t=-pi/2:0.1:pi/2;
```

Change to the Icon page, and set **Icon frame** to **Invisible** and **Drawing coordinates** to **Normalized**. Place the following commands in **Drawing commands**:

```
plot([0.5,0,0,0.5],[0,0,1,1],0.5+0.5*cos(t),0.5+0.5*sin(t))
text(0.05,0.65,'a');
text(0.05,0.2,'b');
text(0.75,0.45,'ab');
```

To produce the OR gate block, the process is similar, using the following in **Drawing commands**:

```
plot([0,0],[0,1],t,0.5*t.^2,t,1-0.5*t.^2);
text(0.05,0.65,'a');
text(0.05,0.2,'b');
text(0.6,0.45,'a+b');
```

The logic gate blocks that result are shown in Figure 6-26.

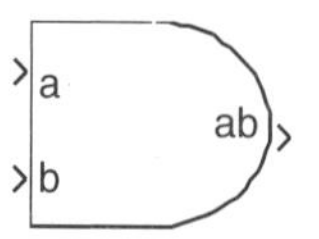

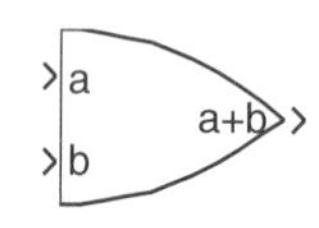

Figure 6-26 Logic gate masked blocks

The final drawing commands, `dpoly(num,denom)` and `droots (zeros, poles, gain)`, display a transfer function on the block's icon.

`dpoly` displays the transfer function in polynomial form. The arguments *num* and *denom* are vectors containing the coefficients of the transfer function numerator and denominator in descending powers of s. `dpoly` will also display discrete transfer functions in either descending powers of z or ascending powers of $1/z$. To display the transfer function in descending powers of z, use the command `dpoly(num,denom,'z')`. To display the transfer function in ascending powers of $1/z$, use `dpoly(num,denom,'z-')`.

`droots` displays a transfer function in factored pole-zero form. *zeros* is a vector containing the zeros of the transfer function (roots of the numerator), and *poles* is a vector containing the poles of the transfer function (roots or the denominator). *gain* is a scalar.

6.3.5 Looking Under and Removing Masks

There are two additional masking commands that permit you to view a subsystem underlying a mask and to delete the mask.

To examine the subsystem underlying a masked block, select the masked block and then choose **Edit:Look under mask**.

To convert a masked block into an unmasked block or subsystem, select the block and then open the Mask Editor. Press the **Unmask** button at the bottom of the Mask Editor. If you change your mind about removing the mask, select the block and choose **Edit:Create Mask**. The masking information will be preserved until you close the model. Once you close the model after removing the mask, it is not possible to restore the mask.

6.3.6 Using Masked Blocks

Once you have created a masked block, it can be copied to a model window in a manner identical to that used to copy a block from a SIMULINK block library. For example, to build the model shown in Figure 6-10a, open a new model window and then drag three copies of the Spring-mass block to the new model window. Add the Constant and Scope blocks and connect the blocks with signal lines as shown. Configure the blocks as shown in Table 6-2.

Table 6-2 Cart model configuration parameters

Field	Spring-mass 1	Spring-mass 2	Spring-mass 3
Left spring constant	k1	k2	k3
Right spring constant	k2	k3	0
Mass	m1	m2	m3
Initial position	1	0	0
Initial velocity	0	0	0

Now the model is complete. Configure the Scope block and simulation parameters as in Example 6-1 and then save the model. In the MATLAB workspace, run the script M-file (`set_x4a.m`) to assign values to the spring constants and masses. This model will produce results identical to the model in Example 6-1.

6.3.7 Creating a Block Library

Each of the block libraries in the SIMULINK block library is a masked subsystem containing a number of SIMULINK blocks that are not connected with signal lines. To verify this, select the Sources block library in the SIMULINK block library and then choose **Edit:Edit Mask** in the SIMULINK block library menu bar. The SIMULINK block library is itself a SIMULINK model containing several masked subsystems, none of which have inputs or outputs.

You can create your own custom block libraries. The simplest way to do this is to copy blocks into an empty model window and then save the model. To use the library, open it using **File:Open** from the SIMULINK block library menu bar, or enter the name of your block library as a command in the MATLAB workspace.

You can also create a subsystem block library by creating a subsystem containing one or more blocks that aren't connected with signal lines. A subsystem library can be masked, as are the block libraries in the SIMULINK block library. The masked subsystem can have a custom icon, but it must not have any dialog box fields (the prompt section of the Initialization page should be empty), and the **Block description field** on the Documentation page must be blank. Otherwise, double clicking on the block library icon would open a block dialog box, instead of opening a subsystem window from which to copy blocks.

Example 6-10

In this example we will create a custom block library. The library will contain a Gain block and a block library containing the Spring-mass block.

Open a new model window.

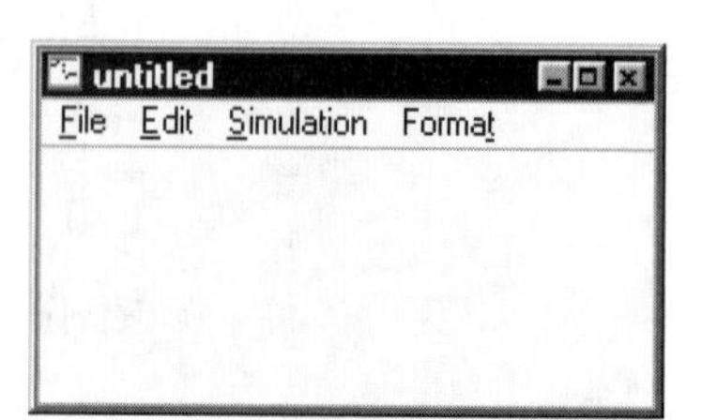

Copy a Gain block from the Linear block library and a Subsystem block from the Connections block library.

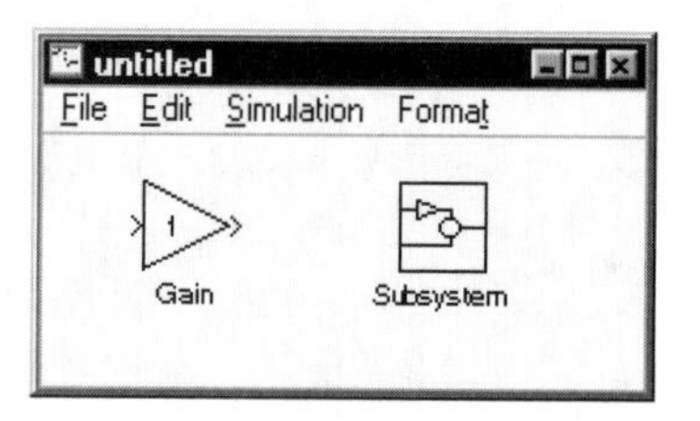

Label the subsystem `My Library`.

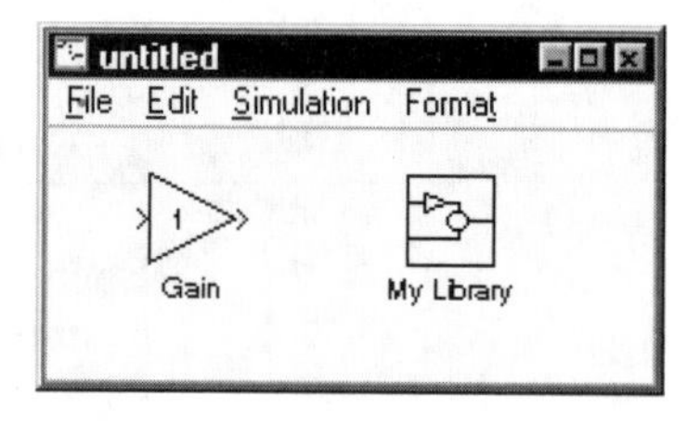

Double click on subsystem My Library, opening the subsystem window. Drag a copy of the Spring-mass block to the subsystem window, and rename the block `Spring-mass`. Then close the subsystem window.

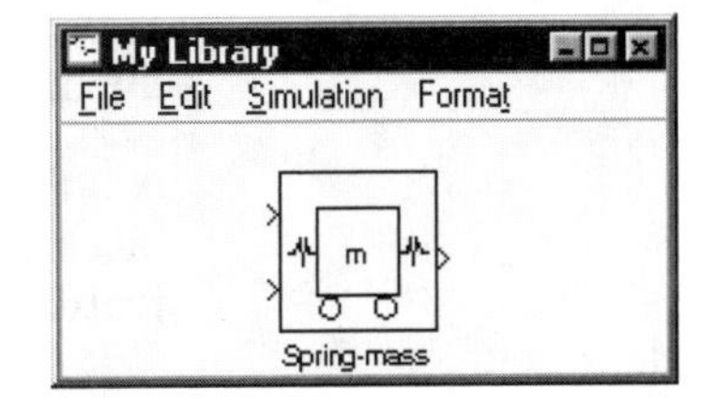

Select the subsystem and choose Edit:Create Mask from the model window menu bar.

In the **Drawing commands** field on the Icon page, enter `disp('Personal\nMasked\nBlocks')`. Leave all other fields in the Mask Editor blank.

Choose **Apply** and **Close**. Resize `My Library` to fit the block icon. Save the model using the name `pers_lib`.

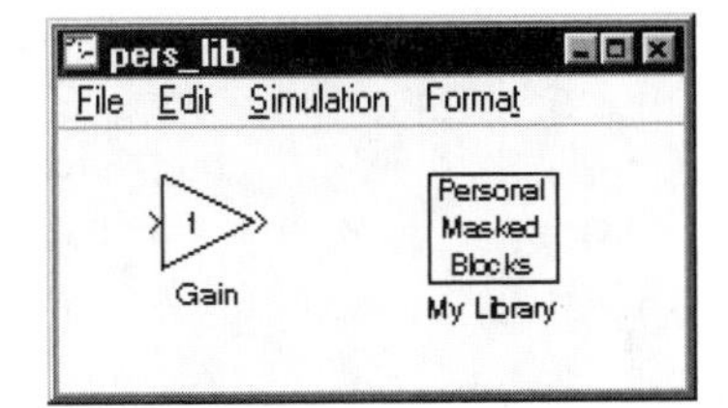

6.4 Conditionally Executed Subsystems

The Connections block library provides two blocks that cause subsystems to execute conditionally. The Enabled subsystem block causes a subsystem to execute only if a control input is positive. The Triggered subsystem block causes a subsystem to execute once when a trigger signal is received. Placing both the Enabled subsystem and Triggered subsystem blocks in a subsystem causes the subsystem to execute once when a trigger signal is received only if an enable input is positive.

6.4.1 Enabled Subsystems

Enabled subsystems permit us to model systems that have multiple operating modes or phases. For example, the aerodynamics of a jet fighter in the landing configuration are quite different from the aerodynamics of the same airplane in supersonic flight. The digital flight control system for such an airplane will likely employ different control algorithms in the different flight regimes. A SIMULINK model of the airplane and its control system might need to include both flight regimes. It is possible to model such a system using logic blocks or switch blocks. If we use this approach, every block in the model will be evaluated each simulation time step, including the blocks that are not currently contributing to the system's behavior. If we convert the various flight dynamics and control algorithm subsystems into enabled subsystems, only the subsystems that are active during a particular simulation step will be evaluated during that step. This can provide a significant computational savings.

A subsystem is converted into an enabled subsystem by adding an Enable block from the Connections block library to the subsystem. Figure 6-27 illustrates a simple proportional controller converted into an enabled subsystem.

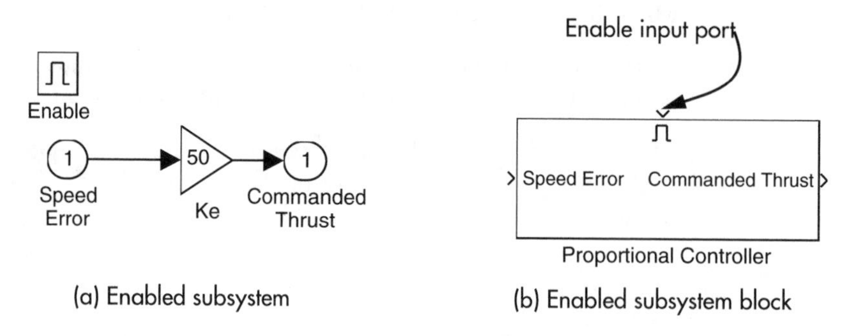

Figure 6-27 Creating an enabled subsystem

The Enable block's dialog box is shown in Figure 6-28. The dialog box has two fields. The first field, **States when enabling**, is a drop-down menu with two options: **reset** and **held**. Choose **reset** to cause any internal states in the subsystem to be reset to the specified initial conditions each time the block is enabled. If you choose **held,** when the block is reenabled, it will resume with all internal states at the values they held when the block was last executed. The second field, **Show output port**, is a check box. When selected, the Enable block will have an output port. This output port passes through the signal received at the Enable input port when the block is enabled.

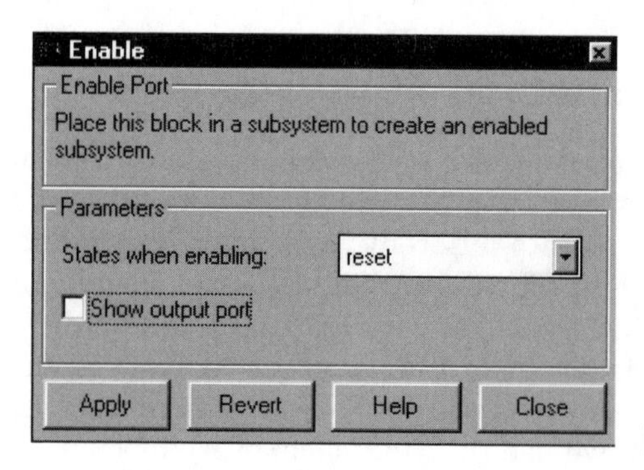

Figure 6-28 Enable block dialog box

It is also important to configure the Outport blocks of an enabled subsystem. The dialog box for the Outport block (Figure 6-29) has three fields. The first field, **Port number**, determines the order in which the ports are displayed on the subsystem block icon. The second field, **Output when disabled**, is a drop-down menu with two options: **reset** and **held**. Choose reset to cause the output to reset to the value in the third field, **Initial output**. Choose **held** to cause the output to remain at the last value output before the subsystem was disabled.

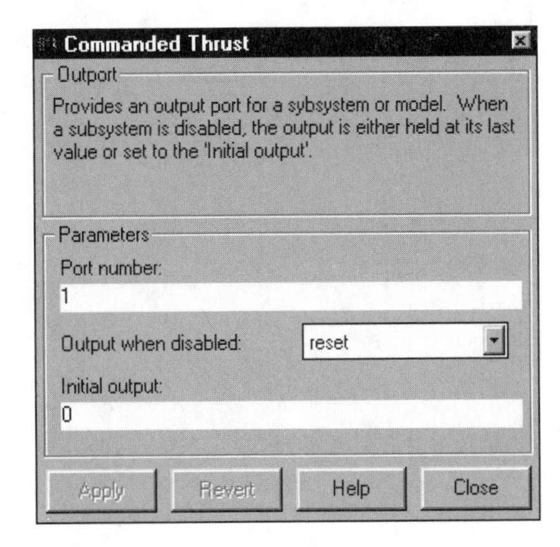

Figure 6-29 Enabled subsystem Outport block dialog box.

An enabled subsystem is enabled when the signal at the Enable input port is positive. The input signal may be either a scalar or vector. If the signal is a vector, the subsystem is enabled if any element of the vector is positive.

Example 6-11

To illustrate the use of enabled subsystems, suppose we wish to modify the automobile speed control of Example 4-6 such that it has two modes of operation depending on the speed error. If the absolute value of speed error

$$v_{err} = |\dot{x}_{desired} - \dot{x}|$$

is less than a threshold value, say 2 ft/s, and the absolute value of the rate of change speed error, $\dot{v}_{err}$, is also less than a threshold value, say 1 ft/s^2, we wish to switch to proportional-integral (PI) control. Once PI control is enabled, it is to remain enabled as long as v_{err} is less than a larger threshold value, 5 ft/s. Otherwise, proportional control is to be enabled.

The SIMULINK model is shown in Figure 6-30. The mode selector subsystem, shown in Figure 6-31, produces two outputs. `Choose PI` is set to `1.0` if the conditions for PI control are satisfied, and `0.0` otherwise. `Choose P` is always the logical inverse of `Choose PI`.

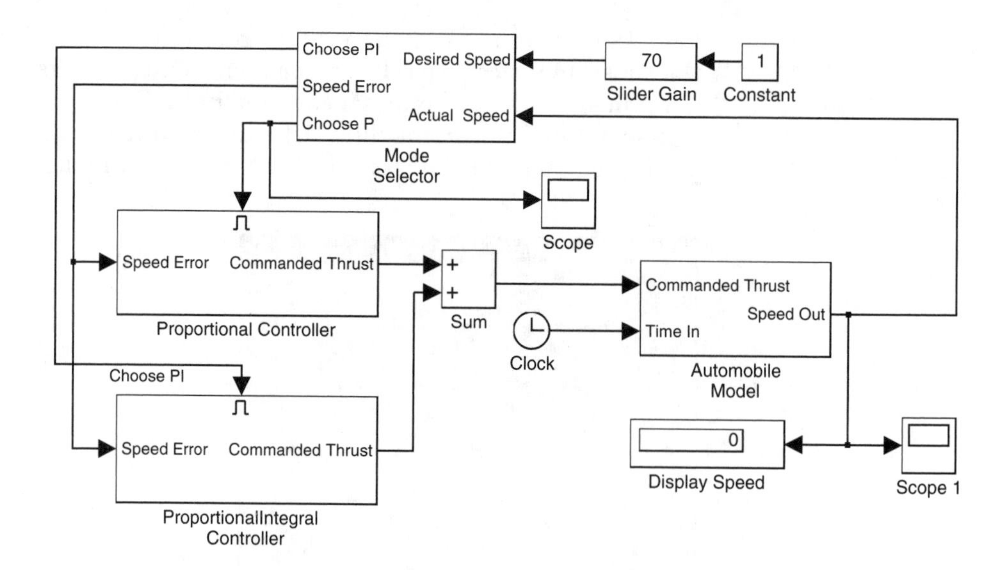

Figure 6-30 SIMULINK model of automobile with a dual-mode speed control

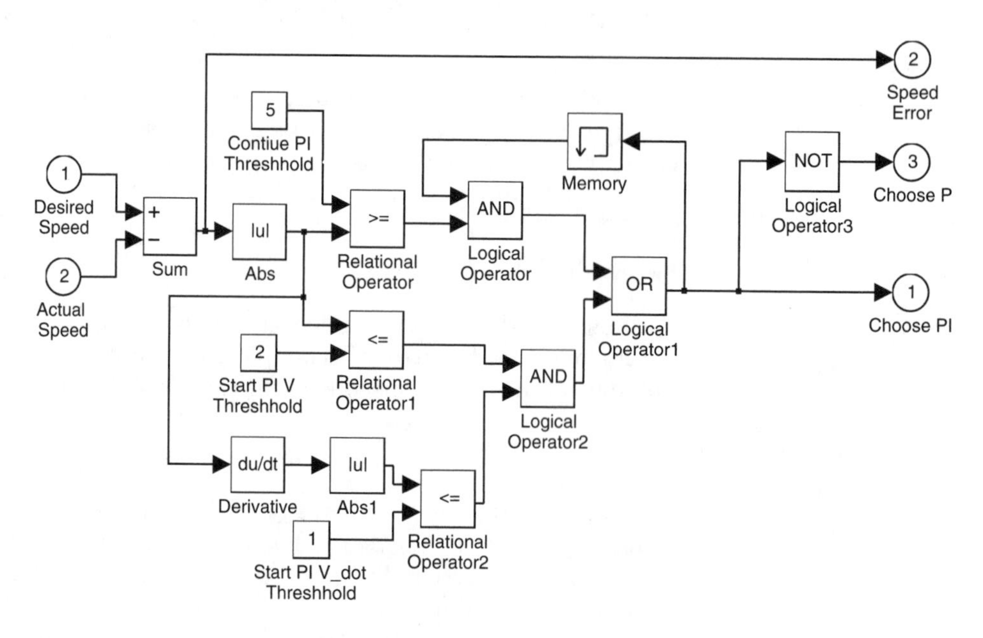

Figure 6-31 Mode selector subsystem

The proportional controller subsystem is illustrated in Figure 6-27. The PI subsystem block is shown in Figure 6-32. The Enable block and Outport block for both controller subsystems are configured to **reset**.

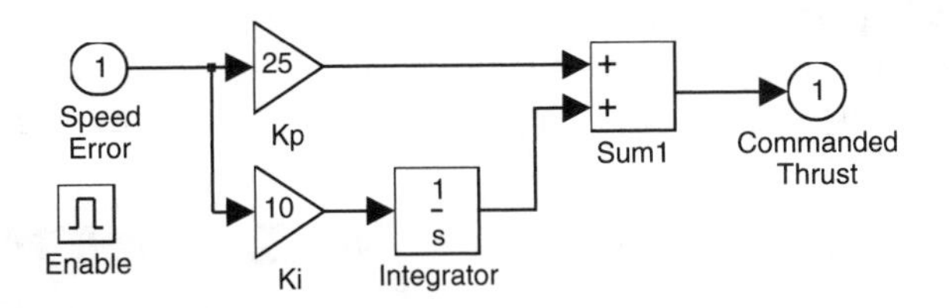

Figure 6-32 Proportional-integral enabled subsystem

Executing the simulation results in the speed trajectory shown in Figure 6-33a. The `Choose P` signal for this simulation is plotted in Figure 6-33b. Initially, proportional control is enabled. Since the rate of change of speed error is less that the threshold value, as soon as the speed error decreases below 2 ft/s, PI control is enabled.

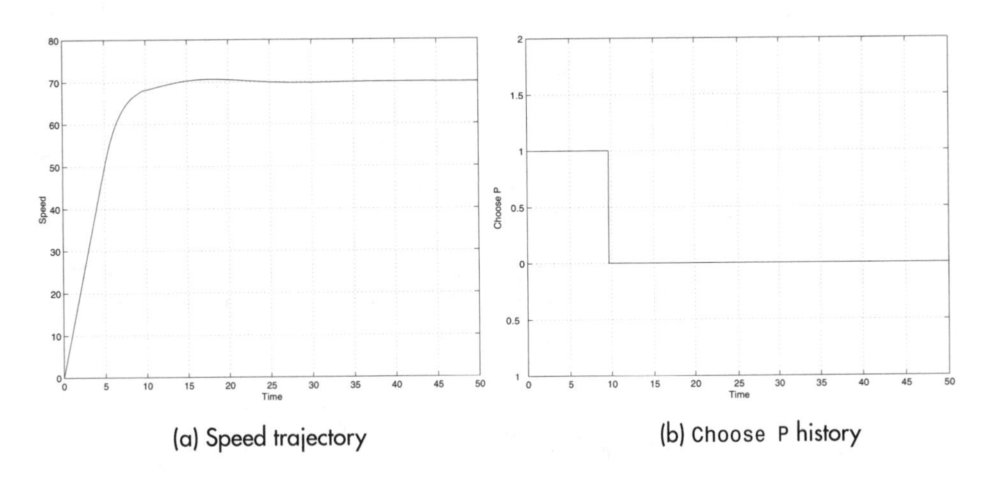

(a) Speed trajectory

(b) `Choose P` history

Figure 6-33 Speed trajectory and `Choose P` signal history for dual mode controller model

6.4.2 Triggered Subsystems

A triggered subsystem is executed once each time a trigger signal is received. A triggered subsystem and the Trigger dialog box are shown in Figure 6-34. The first field, **Trigger type**, is a drop-down menu with three choices: **rising**,

falling, and **either**. If **rising** is selected, a trigger signal is defined as the trigger input crossing zero while increasing. If **falling** is selected, the trigger signal is defined as the trigger input crossing zero decreasing. If **either** is selected, the trigger signal is defined as the trigger input crossing zero increasing or decreasing. The second field in the Trigger block dialog box, **Show output port**, is a check box. If the box is checked, the Trigger block will have an output port that passes through the trigger signal.

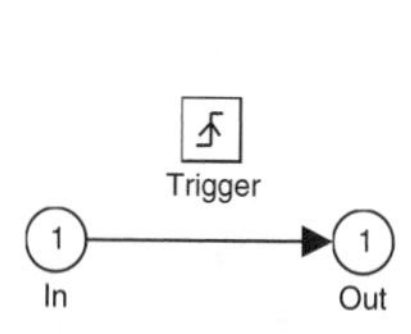

(a) Triggered subsystem

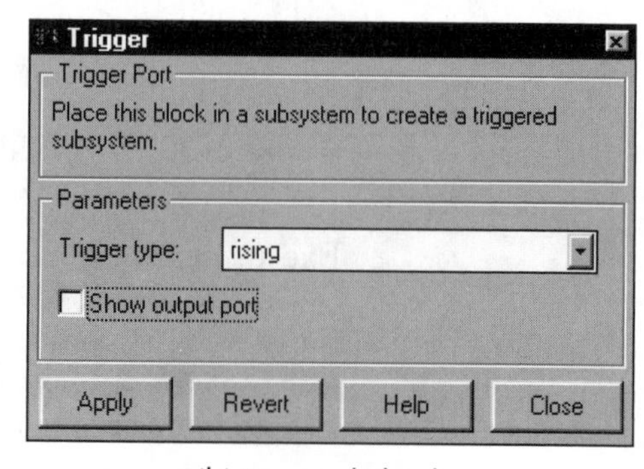

(b) Trigger dialog box

Figure 6-34 Triggered subsystem

A triggered subsystem holds its output value after the trigger signal is received. The initial output value of a triggered subsystem is set using the subsystem's Outport blocks.

The trigger signal may be either a scalar or a vector. If the signal is a vector, the subsystem is triggered when any element of the vector satisfies the **Trigger type** selection.

Example 6-12

The triggered subsystem illustrated in Figure 6-34 passes its input to its output when a trigger signal is received, and is configured (see Output block) to hold its output at that value until another trigger signal is received. The Outport block is initialized to 0. In Figure 6-35 we have added this subsystem to the automobile speed control model and routed the subsystem output to a Display block from the Sinks block library. The trigger input is connected to the enable signal for the PI controller. The Display block displays 0 until the PI controller is activated, and afterward, it displays the most recent activation time.

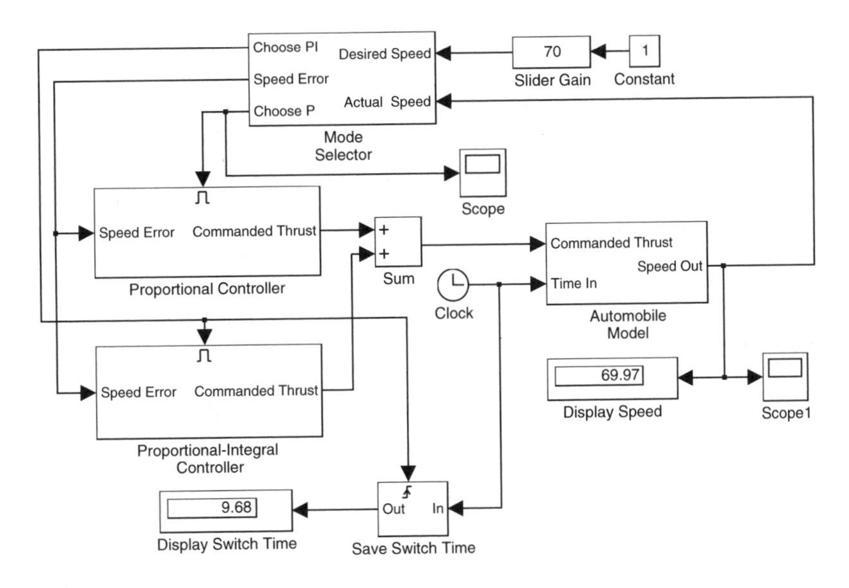

Figure 6-35 Automobile model with triggered subsystem

6.4.3 Triggered and Enabled Subsystems

Placing both an Enable block and a Trigger block in a subsystem produces a triggered and enabled subsystem. This subsystem will have both a trigger input and an enable input. The subsystem behaves the same as a triggered subsystem, but the trigger signal is ignored unless the enable signal is positive.

6.4.4 Discrete Conditionally Executed Subsystems

All three types of conditionally executed subsystems may contain continuous blocks, discrete blocks, or both continuous and discrete blocks. Discrete blocks in an enabled subsystem execute based on their sample time. They use the same time reference as the rest of the SIMULINK model; time in the subsystem is referenced to the start of the simulation, and not the activation of the subsystem. Consequently, an output dependent on a discrete block won't necessarily change at the instant the subsystem is enabled.

Discrete blocks in triggered subsystems must have their sample times set to –1, indicating that they inherit their sample time from the driving signal. Note that discrete blocks that include time delays (z^{-1}) change state once each time the subsystem is triggered.

6.5 Summary

In this chapter we have described several SIMULINK features that make it practical to model complex systems. We discussed SIMULINK subsystems, which provide a facility similar to subprograms in traditional programming languages. We then discussed masking, which allows us to create subsystems that hide their functionality. Finally, we discussed conditionally executed subsystems.

6.6 Reference

1 Hanselman, Duane, and Littlefield, Bruce, *The Student Edition of MATLAB Version 5 User's Guide*. Upper Saddle River, N.J.: Prentice-Hall, 1997, pp. 93–102.

7

SIMULINK Analysis Tools

In this chapter, you'll learn to use SIMULINK analysis tools to gain understanding of SIMULINK models and to aid in the design process. Using the linearization tools we'll extract linear state-space models from block diagrams. We'll then use the trim tools to find equilibrium points.

7.1 Introduction

In the previous chapters, we have described the process of modeling dynamical systems using SIMULINK. In this chapter, we will show how you can use SIMULINK to gain insight into the behavior of those systems.

All of the analysis and design capabilities we will discuss in this chapter are used from within the MATLAB workspace. From the MATLAB prompt, you can determine the structure of a model's state vector, the number of inputs and outputs, and other important parameters. You can run the simulation from within the MATLAB workspace using the MATLAB function `sim`, and can change certain model parameters and inputs. The linearization commands (`linmod`, `linmod2`, and `dlinmod`) allow you to linearize a SIMULINK model about any point in its state space. The trim command (`trim`) locates equilibrium points. These tools can be used both to analyze the behavior of a system and to facilitate the design of certain system parameters.

7.2 Determining Model Characteristics

To use many of the capabilities of the analysis tools to be discussed in this chapter, you must know the structure of the SIMULINK state vector for the model. In this section we will start with a brief discussion of SIMULINK state vectors. Then we will show how to determine the structure of a model's state vector from the MATLAB command line.

7.2.1 SIMULINK State Vector Definition

A SIMULINK model is a graphical description of a set of differential and difference equations. SIMULINK converts this graphical representation into a state-space representation consisting of a set of simultaneous first-order differential and difference equations. For example, the second-order continuous system shown in Figure 7-1 is represented by SIMULINK as two first-order differential equations as follows:

$$\begin{aligned}\dot{x}_1 &= x_2 \\ \dot{x}_2 &= -0.5x_1 - 2x_2.\end{aligned} \tag{7-1}$$

Here, x_1 is a state variable corresponding to the output of the Integrator block labeled `Displacement`, and x_2 is a state variable corresponding to the output of the Integrator block labeled `Velocity`. Recall that the state variable representation of a system is not unique. An equally valid choice of state variables in this example would be to associate x_1 with `Velocity` and x_2 with `Displacement`. (Of course, if we change the state variable definition, we also must write the differential equations in terms of the new state variables.)

Since the analysis tools we will discuss in this chapter work in terms of the SIMULINK state vector, we must know how SIMULINK structures the state vector for a particular model. For example, the `sim` command includes the option to set the initial conditions for the model's state vector. If you wish to set the initial value of velocity to 1.0 and the initial value of displacement to 0.0, you must know which state variable corresponds to each quantity.

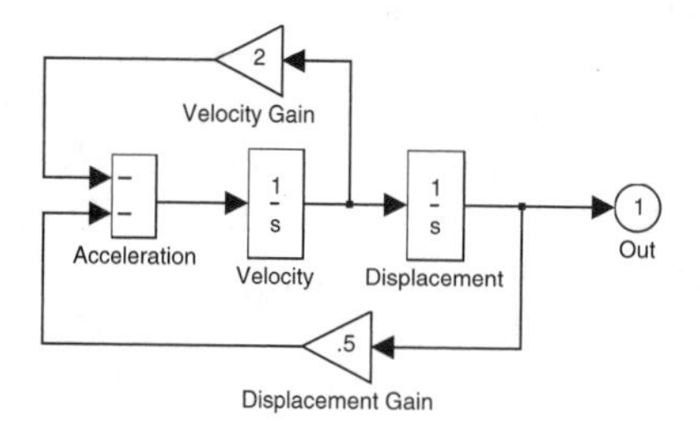

Figure 7-1 SIMULINK model of second-order continuous system

SIMULINK divides a model's state vector into continuous and discrete components. There is one continuous state vector component for each continuous integrator, including integrators present implicitly, due, for example, to Transfer Fcn blocks and State-Space blocks. Similarly, there is a discrete state vector component for each time delay, including delays present implicitly.

7.2.2 Using the `model` Command

The *`model`* command permits us to determine the structure of a SIMULINK model's state vector. There are three versions of the *`model`* command as follows:

```
sizes = model([],[],[],0)
[sizes, x0] = model([],[],[],0)
[sizes,x0,states] = model([],[],[],0)
```

where *model* is the name of the SIMULINK model. Note that the input arguments are the same for each version. The first three arguments are empty matrixes, and the fourth argument is 0.

The first output argument, `sizes`, is a six-element vector defined in Table 7-1. The first two elements of `sizes` contain the number of continuous and discrete states. So, for example, if `sizes(1)` is 2 and `sizes(2)` is 3, the state vector will have a total of five elements. The first two elements will be the continuous states, and the last three elements will be the discrete states.

Table 7-1 Contents of the `sizes` vector

Component	Meaning
`sizes(1)`	Number of continuous states. This will include explicit states associated with Integrator blocks, and implicit states, such as those associated with Transfer Fcn blocks and State-Space blocks.
`sizes(2)`	Number of discrete states. This will include explicit states associated with Unit Delay blocks and implicit states, such as those associated with Discrete Transfer Fcn blocks.
`sizes(3)`	Number of outputs. An output is counted for each component of each Outport block. Thus if the signal entering an Outport block is a scalar, there will be one output associated with that block. If the signal entering an Outport block is a three-component vector, there will be three outputs associated with that block. Note that To Workspace blocks and To File blocks do not count as outputs.
`sizes(4)`	Number of inputs. An input is counted for each component of each Inport block. Note that From Workspace blocks and From File blocks do not count as inputs.
`sizes(5)`	Number of discontinuous roots in the system. This information does not pertain to the analysis tools discussed in this chapter.
`sizes(6)`	Flag that is set to 1 if a subsystem has direct feedthrough of an input. This information does not pertain to the analysis tools discussed in this chapter.

The second output argument (`x0`) is optional. If present, it is the initial value of the SIMULINK model's state vector. Recall that the integrator states may be initialized using the dialog box for each integrator, and that this initialization may be overridden using the Workspace I/O page of the **Simulation:Parameters** dialog box.

The third output argument (`states`) is a cell array that identifies the block associated with each component of the SIMULINK state vector using the convention

```
model file name/top level subsystem/2nd level subsystem/.../block
```

So, for example, the state associated with an Integrator block labeled `Velocity`, at the top level of a model named `sysmdl_a`, would be `sysmdl_a/Velocity`. If this same block were located in a subsystem named `subsys_1` in model `sysmdl_a`, the state would be named `sysmdl_a/subsys_1/Velocity`.

Example 7-1

Consider the SIMULINK model shown in Figure 7-2. Suppose we have saved this model in a file named `sysmdl_a.mdl`.

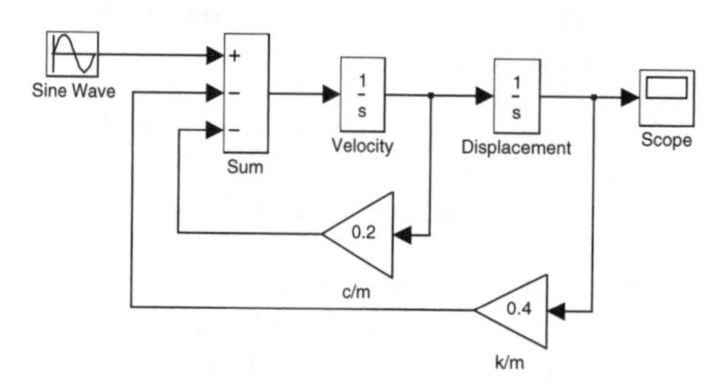

Figure 7-2 Second-order system with no inputs or outputs, stored in `sysmdl_a.mdl`

From the MATLAB prompt, we enter the following command:

```
[sizes, x0, states]=sysmdl_a([],[],[],0)
```

resulting in the output shown in Figure 7-3.

From this output we can determine the structure of the model state vector. `sizes(1)` is 2, so there are two continuous states, as we would expect since the model has two Integrator blocks. `sizes(2)` is 0, so there are no discrete states. Using `sizes` and `states`, the state vector can be described as follows:

Component	Type	Associated with block
1	Continuous	Displacement
2	Continuous	Velocity

```
sizes =
      2
      0
      0
      0
      0
      0
      1
x0 =
      0
      0
      0
      0
states =
    'sysmdl_a/Displacement'
    'sysmdl_a/Velocity'
```

Figure 7-3 Model characteristics of `sysmdl_a`

Example 7-2

We can also identify the model characteristics for a SIMULINK model containing subsystems. Suppose we saved the model shown in Figure 7-4 in file `sysmdl_b.mdl`. From the MATLAB prompt, we enter the command

```
[sizes, x0, states]=sysmdl_b([],[],[],0)
```

resulting in the output shown in Figure 7-5.

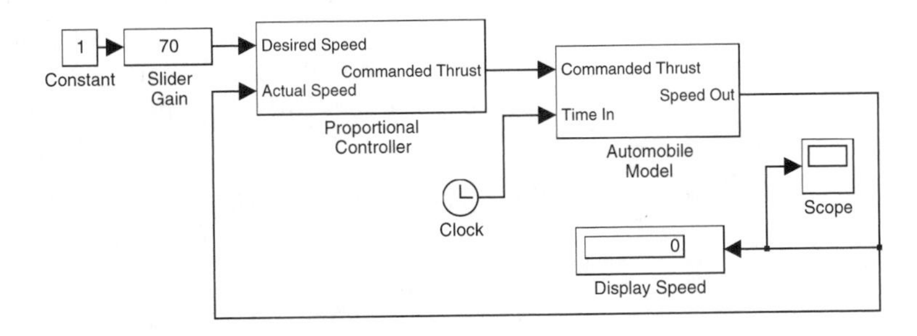

Figure 7-4 Automobile model using subsystems

```
sizes =
     2
     0
     0
     0
     0
     0
     1
x0 =
     0
     0
     0
     0
states =
    'sysmdl_b/Automobile Model/Speed'
    'sysmdl_b/Automobile Model/Distance'
```

Figure 7-5 Model characteristics of `sysmdl_b`

Note that the block names in `states` consist of the full path to the block, including the name of the subsystem.

7.3 Executing Models from MATLAB

The MATLAB `sim` command permits us to run SIMULINK models from the MATLAB prompt or from within M-files. This capability makes it easy to access model states for analysis or to run a simulation repeatedly using different values for parameters, different inputs, or different initial conditions. `sim` may be used in conjunction with MATLAB command `simset`, which creates and edits a simulation options data structure, and `simget`, which gets the simulation options structure from a model. After discussing these commands, we will show how they can be used to facilitate the analysis of model behavior.

7.3.1 Using `sim` to Run a Simulation

The syntax of the `sim` command is

```
[t,x,y] = sim(model,TimeSpan,Options,ut)
```

The return variable list is optional. It can be omitted or can contain only one return variable, two return variables, or all three return variables. The first returned variable (`t`) contains the value of simulation time at each output point. Recall that the default behavior is to produce an output point at the end of each integration step, but that the Output options section of the **Simulation:Parameters** dialog box allows you to override this default either by adding output points or explicitly specifying the time of each output point. The second returned variable (`x`) is the state variable trajectory, with one column for each state variable and one row corresponding to each time point in the first returned variable (`t`). The final returned variable (`y`) contains the output variable trajectory, one column for each output vector element, and one row corresponding to each time point. If there are no Outport blocks, but an output variable (`y`) is present, it will contain the empty vector `[]`.

The first argument to the sim command (*`model`*) is a MATLAB string containing the name of the SIMULINK model, without the file name extension (`.mdl`). This argument is mandatory. The remaining arguments are optional, and allow us to override various model configuration parameters.

`TimeSpan` specifies the output time points. If `TimeSpan` is specified, it overrides the output times specified in the **Simulation:Parameters** dialog box. `TimeSpan` may have four different forms, as listed in Table 7-2.

Table 7-2 Specifying the simulation output times in the `sim` command

Value of TimeSpan	Output time point values
`[]`	Default to values specified in the **Simulation:Parameters** dialog box.
`[T_Final]`	Time points default to values specified in the **Simulation:Parameters** dialog box. The simulation will stop when the simulation time reaches `T_Final`.
`[T_Start T_Final]`	The simulation will start at time `T_Start` and stop at `T_Final`. The value of time at intermediate points will be as specified in the **Simulation:Parameters** dialog box. Note that this is a two-element vector.
`[OutputTimes]`	If `OutputTimes` is a vector of three or more components, there will be one output time point corresponding to each element of `OutputTimes`. The contents of the first return variable (`t`) will be identical to `OutputTimes`. Probably the most common form of `OutputTimes` is `[T_Start:TimeSpacing:T_final]`, producing a vector of equally spaced points.

`Options` is a MATLAB data structure that allows you to override many of the parameters in the **Simulation:Parameters** dialog box. The options structure is created, updated, and displayed using the MATLAB command `simset`, which we will discuss shortly.

`ut` overrides the Load from workspace section of the Workspace I/O page of the **Simulation:Parameters** dialog box. `ut` may be either an input table or a string containing the name of a MATLAB function. If `ut` is an input table, it must be of the form `[t,u1,u2,...]`, where `t` is a column vector of time points, and `u1`, `u2`, etc. are column vectors of the values of the inputs corresponding to the time points in `t`. The inputs are associated with Inport blocks, ordered according to the Inport block numbers. SIMULINK treats `ut` as a lookup table, linearly interpolating between time points. If `ut` is a string, it must name a function that returns an input vector, given a single argument of time.

When input arguments are supplied to the `sim` command to override simulation parameters, the SIMULINK model is not changed. The `sim` command arguments affect the SIMULINK model only during the execution of the simulation. When the simulation stops, the model parameters are the same as they were before the `sim` command was executed.

Example 7-3

Suppose we wish to set `ut` for a model with a single input according to the rule

$$u = \frac{1}{2}t \qquad t \le 1$$
$$u = t - \frac{1}{2} \qquad t > 1$$

The following MATLAB statement will create an input table compatible with the `sim` command. SIMULINK will interpolate in this table, producing values that correspond to the rule.

```
ut=[[0,1,100]',[0,1/2,99.5]']
```

Example 7-4

Suppose we wish to set `ut` for a model with two inputs to the vector function

$$\boldsymbol{u}(t) = \begin{bmatrix} \sin t & \cos t \end{bmatrix}$$

Figure 7-6 illustrates a function M-file that returns this vector.

```
function u = ut_fun(t)
u = [sin(t),cos(t)] ;
```

Figure 7-6 M-file to return the vector [sin(t),cos(t)]

We set `ut` using the MATLAB statement

```
ut='ut_fun'
```

Example 7-5

In this example we run the model illustrated in Figure 7-2 using the model's default configuration. This is equivalent to selecting **Simulation:Start** from the model's menu bar. At the MATLAB prompt, enter the command

```
sim('sysmdl_a')
```

SIMULINK will run the model and, if the Scope is open, display the plot of the output when the simulation is complete.

7.3.2 Setting Simulation Parameters with `simset`

The `simset` command is used to create and edit the options structure. There are three forms of the `simset` command. The first form is

```
options = simset('name_1',value_1,'name_2',value_2,...)
```

where *name_1*, `value_1`, etc. are property name, property value pairs as defined in Table 7-3. Refer to Section 3.8 for a detailed description of the simulation parameters. Notice that the property names are MATLAB strings. The property values may be either numbers or strings. So, for example, to assign the value of 0.0001 to the relative error tolerance and to select the fourth-order fixed-step-size Runge-Kutta solver, you could use the following MATLAB command:

```
opts = simset('RelTol',1.0E-4,'Solver','ode4')
```

The second form of `simset` is

```
options = simset(oldopts,'name_1',value_1,...)
```

which allows you to change an existing options structure, either by changing previously set options or by adding more options.

The third form of simset is

```
options = simset(oldopts,newopts)
```

where both oldopts and newopts are options structures previously created using simset. This form of simset merges the contents of oldopts and newopts, with the contents of newopts having priority.

Table 7-3 describes the properties that may be set using simset.

Table 7-3 Simulation property names and values

Property name	Property value and associated Simulation:Parameters fields
Solver	Select the SIMULINK differential equation solver. Permissible values are 'ode45', 'ode23', 'ode113', 'ode15s', 'ode23s', 'ode5', 'ode4', 'ode3', 'ode2', 'ode1', 'FixedStepDiscrete', 'VariableStepDiscrete'
RelTol	Override the contents of the **Relative tolerance** field.
AbsTol	Override the contents of the **Absolute tolerance** field.
Refine	Equivalent to setting **Output options** to **Refine output**. The property value must be set to a positive integer. If the output time points are specified explicitly, this property is ignored.
MaxStep	Upper bound on integration step size. Overrides **Max step size**.
Initial-Step	Initial integration step size. Overrides the contents of **Initial step size**.
MaxOrder	Maximum order if solver ODE15S is used. Otherwise ignored.
FixedStep	Integration step size if a fixed step solver is used. Overrides **Fixed step size**.
Output-Points	Set to either the string 'specified' or the string 'all'. Select 'specified' (the default) to produce output points only at the time points specified in TimeSpan. Select 'all' to produce output points at the time points specified in TimeSpan and at each integration step. This property is equivalent to selecting **Produce specified output only** or **Produce additional output** in **Output options**.

Table 7-3 Simulation property names and values (Continued)

Property name	Property value and associated Simulation:Parameters fields
`Output-Variables`	Overrides the three check boxes in the Save to workspace section of the Workspace I/O page. The property value is a string of up to three characters. If the string contains 't', simulation time is output to the MATLAB workspace, equivalent to selecting the **Time** check box. Including 'x' is equivalent to selecting the **States** check box, and including 'y' is equivalent to selecting the **Output** check box. Thus this property value could be set to 'txy', 'ty', 'yx', etc.
`MaxRows`	Places an upper bound on the number of rows in the output matrixes. Equivalent to selecting the **Limit to last** check box and entering a value in the **Limit to last** input field.
`Decimation`	Overrides the **Decimation** field. Setting this property value to 1 causes every point to be output, setting it to 2 causes every other point to be output, and so on. Must be a positive integer.
`Initial-State`	Set to a vector containing the initial values of all of the model's state variables. Overrides the **Load initial** check box and field in the States section of the Workspace I/O page.
`Final-StateName`	Character string containing the name of the MATLAB workspace variable in which to save the final value of the state vector. Overrides the **Save final** check box and field in the States section of the Workspace I/O page.
`Trace`	Set to contain a comma separated list of strings that may include 'minstep', 'siminfo', 'compile', or ''. For example, to include all three options, the string could be 'siminfo,compile,minstep'. 'minstep' is equivalent to setting the **Minimum step size violation** event to **warning**. 'siminfo' causes SIMULINK to produce a brief report listing key parameters in effect at the start of a simulation. 'compile' causes SIMULINK to produce a model compilation listing in the MATLAB workspace as the model is compiled, before it runs. The compilation listing is intended primarily for use by The MathWorks in troubleshooting SIMULINK problems.

Table 7-3 Simulation property names and values (Continued)

Property name	Property value and associated Simulation:Parameters fields
SrcWork-space	Select the MATLAB workspace in which to evaluate MATLAB expressions defined in the model. May be set to 'base' (the default), 'current', or 'parent'. This option is extremely important if you are running a SIMULINK model from within a function M-file. 'base' indicates that all variables used to initialize block parameters are to be taken from the base MATLAB workspace. 'current' indicates that the variables are to be taken from the private workspace from which sim is called. 'parent' indicates that the variables are to be taken from the workspace from which the current function (the one that called sim) was called (the next higher function). No corresponding field in **Simulation:Parameters**.
DstWork-space	Select the MATLAB workspace in which to assign MATLAB variables defined in the model. May be set to 'base', 'current', or 'parent'. DstWorkspace determines where To Workspace blocks send their results. No corresponding field in **Simulation:Parameters**
ZeroCross	Overrides the **Disable zero crossing detection** check box. May be either 'on' or 'off'.

It is particularly important to ensure that the properties SrcWorkspace and DstWorkspace are set properly. There are three options for each: 'base', 'current', and 'parent'. The base workspace is the MATLAB prompt. Thus variables defined in the base workspace are those variables listed when the command who is entered at the MATLAB prompt. Each function M-file has a private workspace; variables defined in a function M-file are not visible outside the M-file. 'current' indicates that variables are to be taken from or returned to the workspace from which the sim command is run. 'parent' indicates that variables are to be taken from or returned to the workspace from which the currently executing function was called. It is permissible for SrcWorkspace and DstWorkspace to be set to different values. It is not possible for some model parameters to be defined in the base workspace, and others in a function M-file.

7.3.3 Getting Simulation Parameters with simget

The command simget gets either the options structure or the value of a single property. To get the options structure from a model, the syntax is

```
opts = simget(model)
```

To get the value of a single property, the syntax is

```
value = simget(model,property_name)
```

where `model` is a MATLAB string containing the name of the SIMULINK model without the file name extension (`.mdl`), and `property_name` is a string containing one of the names listed in Table 7-3. `property_name` can also be a cell array in which each cell is a string containing one of the names listed in Table 7-3.

Example 7-6

A useful tool in the analysis of nonlinear second-order systems is the *state portrait*. Consider the following system, discussed in detail by Scheinerman [5]:

$$\dot{x}_1 = -x_2$$
$$\dot{x}_2 = x_1 + x_2^3 - 3x_2.$$

A state portrait graphically depicts the behavior of the system in the state plane (the (x_1, x_2) plane) in the vicinity of an *equilibrium*. An equilibrium is a point in the state plane at which both state derivatives are zero. If, in some sufficiently small region in the vicinity of an equilibrium, every trajectory converges to the equilibrium, the equilibrium is said to be stable. This system has one equilibrium, located at the origin. A SIMULINK model of this system is shown in Figure 7-7.

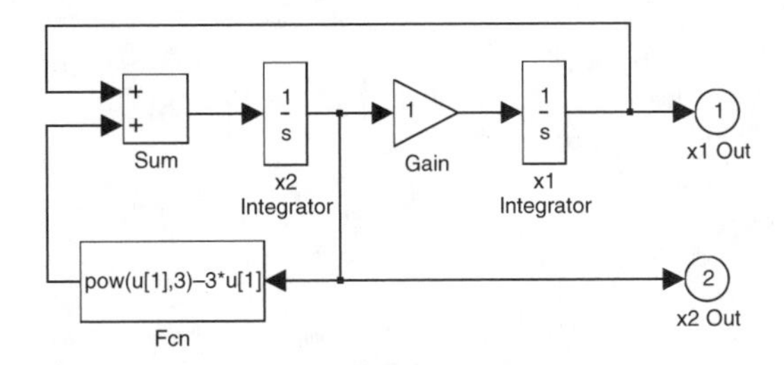

Figure 7-7 SIMULINK model of nonlinear second-order system

We start the analysis process by identifying the elements of the model's state vector. In Figure 7-8 we execute the model, stored in file `sysmdl_e.mdl`, with the fourth argument (`flag`) set to 0. We note that there are two states, the first associated with x_1, the second with x_2.

We can produce a simple state portrait by repeatedly running the model, starting each simulation at a different point in the state space. An M-file that produces a state portrait in this manner is shown in Figure 7-9. The state por-

```
EDU» [states,x0,sizes]=sysmdl_e([],[],[],0)
states =
     2
     0
     2
     0
     0
     0
     1
x0 =
     0
     0
sizes =
    'sysmdl_e/x1 Integrator'
    'sysmdl_e/x2 Integrator'
```

Figure 7-8 Identifying the states from the MATLAB prompt

trait is shown in Figure 7-10. Notice that each state trajectory goes to the origin and stops there. Thus from the state portrait it appears that the origin is a stable equilibrium.

```
% Produce a state portrait for the SIMULINK model sysmdl_e
%
for x1 = -1:0.5:1
  for x2 = -1:0.5:1
    opts = simset('InitialState',[x1,x2]) ;
    [t,x,y] = sim('sysmdl_e',15,opts) ;
    hold on ;
    plot(x(:,1),x(:,2)) ;
  end
end
axis([-1,1,-1,1]);
xlabel('x1') ;
ylabel('x2') ;
grid ;
```

Figure 7-9 M-file to produce simple state portrait

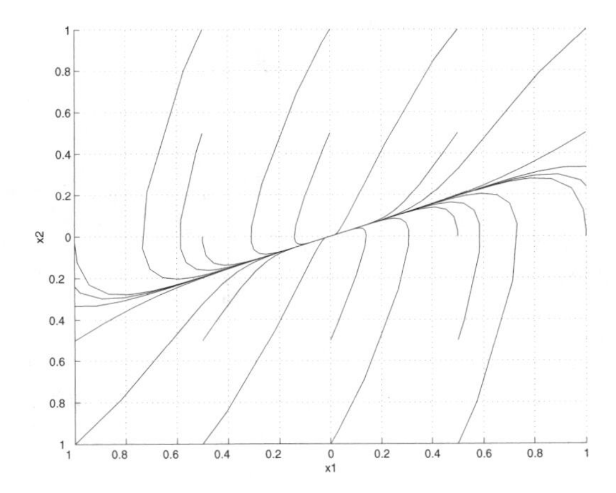

Figure 7-10 Simple state portrait

Example 7-7

The simple state portrait in the previous example shows the system behavior close to the origin and is easy to produce, but there are some problems with this approach. First, there is no easy method to determine the proper `TimeSpan`. Some of the trajectories approach the origin rapidly, others slowly. A more serious problem is that if the simulation is started at a point in the state space where the system is unstable, one or more of the state variables will rapidly diverge, causing floating-point overflow and the failure of the SIMULINK simulation. This is not a weakness of SIMULINK; it is the natural consequence of modeling the behavior of an unstable system.

The M-file illustrated in Figure 7-11 overcomes these problems. As before, we run the simulation repeatedly, starting each simulation at a different point in the state space. However, each simulation performs only one integration step. Using the final point of each state trajectory and the corresponding initial point, we form an approximation of the slope of the state portrait at each initial point. Using these slopes, we produce the quiver plot shown in Figure 7-12. In addition to being easier to read, this state portrait shows the behavior of the system in the unstable region of the state plane.

```
% Produce a state portrait for the SIMULINK model sysmdl_e
% using a quiver plot
h = 0.01 ; % Fixed step size
opts = simset('Solver','ode5','FixedStep',h) ;
x1 = -2.5:0.25:2.5 ;   % Set up the x1 and x2 grid
x2 = -2.5:0.25:2.5 ;
[nr,nc] = size(x1) ;
x1m = zeros(nc,nc) ;  % Allocate space for the output vectors
x2m = x1m ;
for nx1 = 1:nc
  for nx2 = 1:nc
    opts = simset(opts,'InitialState',[x1(nx1),x2(nx2)]) ;
    [t,x,y] = sim('sysmdl_e',h,opts) ;
    dx1 = x(2,1)-x1(nx1) ;
    dx2 = x(2,2)-x2(nx2) ;
    l = sqrt(dx1^2 + dx2^2)*7.5 ;  % Scale the arrows
    if l > 1.e-10
      x1m(nx2,nx1)=dx1/l ;         % Notice the reversed indexes
      x2m(nx2,nx1)=dx2/l ;
    end
  end
end
quiver(x1,x2,x1m,x2m,0) ;
axis([-2.5,2.5,-2.5,2.5]) ;
xlabel('x1') ;
ylabel('x2') ;
grid ;
```

Figure 7-11 M-file to produce quiver plot state portrait

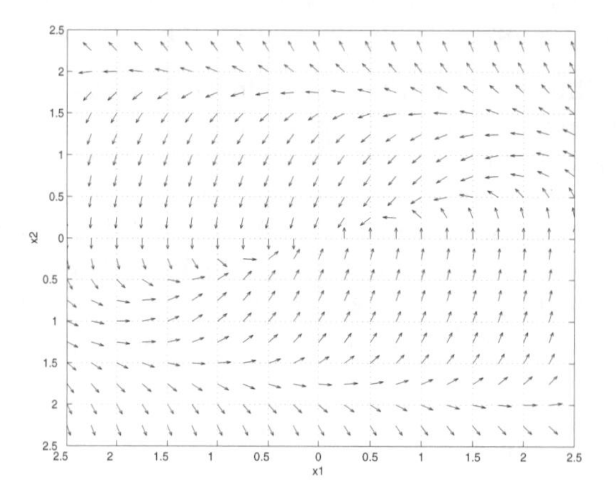

Figure 7-12 Quiver plot state portrait

7.4 Linearization Tools

While the dynamics of most physical systems are nonlinear, many useful techniques for analysis and control system design rely on linear models. For example, frequency domain analysis tools, such as Bode plots and root locus plots, are based on linear systems theory. Modern control techniques, such as pole placement via state variable feedback and the linear quadratic regulator, also rely on linear systems theory. Thus it is frequently convenient to form linear approximations to nonlinear systems. We will begin this section with a short discussion of linearization. Then we will show how the SIMULINK linearization tools may be used to facilitate analysis of nonlinear systems and control system design.

7.4.1 Linearization

A linear model of a nonlinear system is formed by computing matrixes of the partial derivatives of the system state vector time rate of change and the output vector with respect to the state vector and input vector. As discussed in Chapter 4, the general form of the state-space model of a dynamical system is

$$\dot{\boldsymbol{x}} = \boldsymbol{f}(\boldsymbol{x}, \boldsymbol{u}, t) \tag{7-2}$$

$$\boldsymbol{y} = \boldsymbol{g}(\boldsymbol{x}, \boldsymbol{u}, t) \tag{7-3}$$

where $\dot{\boldsymbol{x}}$, $\boldsymbol{f}$, and $\boldsymbol{x}$ are vectors of order n, $\boldsymbol{u}$ is an input vector of order m, and $\boldsymbol{y}$ and $\boldsymbol{g}$ are vectors of order p. Given a nominal point $\boldsymbol{x}$ and a nominal input $\boldsymbol{u}$, at some time t, we can approximate to first order the change in the j-th component of the state vector time rate of change ($\dot{x}_j$) due to a small variation in the i-th component of the state vector (x_i) as

$$\delta\dot{x}_{ji} = \frac{\partial}{\partial x_i} f_j(\boldsymbol{x}, \boldsymbol{u}, t)\delta x_i \tag{7-4}$$

and the change due to a small variation in the k-th component of the input vector as

$$\delta\dot{x}_{jk} = \frac{\partial}{\partial u_k} f_j(\boldsymbol{x}, \boldsymbol{u}, t)\delta u_k \tag{7-5}$$

The total change in $\dot{\boldsymbol{x}}$ due to a small variation in each element of the state vector and input vector can be approximated by the matrix equation

$$\delta\dot{\boldsymbol{x}} = \boldsymbol{A}\delta\boldsymbol{x} + \boldsymbol{B}\delta\boldsymbol{u} \tag{7-6}$$

where each element of $\boldsymbol{A}$ is

$$A_{ji} = \frac{\partial}{\partial x_i} f_j(\boldsymbol{x}, \boldsymbol{u}, t) \tag{7-7}$$

and each element of $\boldsymbol{B}$ is

$$B_{jk} = \frac{\partial}{\partial u_k} f_j(\boldsymbol{x}, \boldsymbol{u}, t) \tag{7-8}$$

Note that $\boldsymbol{A}$ is composed of n rows and n columns, and $\boldsymbol{B}$ of n rows and m columns.

Using similar reasoning, we can approximate the change in the output vector due to small variations in the state and input as

$$\delta \boldsymbol{y} = \boldsymbol{C}\delta \boldsymbol{x} + \boldsymbol{D}\delta \boldsymbol{u} \tag{7-9}$$

where

$$C_{ji} = \frac{\partial}{\partial x_i} g_j(\boldsymbol{x}, \boldsymbol{u}, t) \tag{7-10}$$

and

$$D_{jk} = \frac{\partial}{\partial u_k} g_j(\boldsymbol{x}, \boldsymbol{u}, t) \tag{7-11}$$

Here $\boldsymbol{C}$ is composed of p rows and n columns, and $\boldsymbol{D}$ is composed of p rows and m columns.

Together, matrixes $\boldsymbol{A}$, $\boldsymbol{B}$, $\boldsymbol{C}$, and $\boldsymbol{D}$ may be used to form the linear state-space approximation

$$\dot{\boldsymbol{x}} = \boldsymbol{A}\boldsymbol{x} + \boldsymbol{B}\boldsymbol{u} \tag{7-12}$$

$$\boldsymbol{y} = \boldsymbol{C}\boldsymbol{x} + \boldsymbol{D}\boldsymbol{u} \tag{7-13}$$

This approximation may be valid only in some small region in the vicinity of the nominal point and nominal input. If the system is time varying, the approximation may be valid for a short period of time. Furthermore, some nonlinear systems can't be effectively linearized because important features of the dynamics are lost in the linearization. However, for a large number of physical systems, the linear approximation is quite useful, providing insight into the system's behavior and providing access to a variety of control system design techniques.

7.4.2 SIMULINK Linearization Commands

SIMULINK provides three commands to extract linear state-space approximations from SIMULINK models. `linmod` forms state-space models of continuous systems. `linmod2` is an alternative to `linmod` that uses different techniques to obtain the linearization, striving to reduce truncation error. `dlinmod` forms linear approximations of systems containing both continuous and discrete components.

The syntax of the linearization commands is as follows:

```
[A,B,C,D]=LINMOD(model,X,U,PARA,XPERT,UPERT)
[A,B,C,D]=LINMOD2(model,X,U,PARA,APERT,BPERT,CPERT,DPERT)
[A,B,C,D]=DLINMOD(model,TS,X,U,PARA,XPERT,UPERT)
```

All of the output arguments and all of the input arguments except *`model`* are optional. If no output arguments are specified, the system matrix (`A`) for the linear model is returned. The input arguments are defined in Table 7-4. The output arguments are the four state-space matrixes, as discussed in Section 7.4.1.

Table 7-4 Linearization command input arguments

Argument	Definition
`model`	MATLAB string containing the name of the SIMULINK model. For example, if the SIMULINK model is stored in file `ex1.mdl`, the first argument would be `'ex1'`. This argument is required.
`X`	Value of the model nominal state vector. SIMULINK will linearize the model about any point in the model's state space. Identify the components of the model's state vector using the procedure described in Section 7.2. `X` defaults to a zero vector of appropriate dimension, if either the argument `X` is not present or if it is the empty matrix `[]`.
`U`	Value of the nominal input vector. SIMULINK will linearize the input matrix and direct the transmittance matrix about the specified input vector and state vector. The components of `U` correspond to the numbering of Inport blocks.

Table 7-4 Linearization command input arguments (Continued)

Argument	Definition
PARA	Vector containing two elements. The first is a perturbation value to be used in computing the numerical partial derivatives, and defaults to 10^{-5} for `linmod` and `dlinmod`. For `linmod2`, `PARA(1)` is the minimum perturbation value, and defaults to 10^{-8}. The second element is the time at which to perform the linearization, and defaults to 0 seconds.
XPERT	Vector of perturbation values for each component of the system state vector. This vector of perturbations overrides the default, computed according to the rule `XPERT=PARA(1)+1e-3*PARA(1)*abs(X)`
UPERT	Vector of perturbation values for each component of the input vector. This vector overrides the default, according to the rule `UPERT=PARA(1)+1e-3*PARA(1)*abs(U)`
APERT	Matrix of maximum perturbation values for the system matrix. `APERT` must be either empty (`[]`), or the same size as `A`. `linmod2` uses a fairly complex algorithm to compute the optimal perturbation value for each element of each matrix. The perturbation value for each element of the system matrix (`A(i,j)`) will be bounded from above by `APERT(i,j)` and from below by `PARA(1)`.
BPERT	Matrix of maximum perturbation values for the input matrix (`B`). The perturbation value for each element of `B` will be bounded from above by the corresponding element of `BPERT` and from below by `PARA(1)`.
CPERT	Matrix of maximum perturbation values for the output matrix (`C`). The perturbation value for each element of `C` will be bounded from above by the corresponding element of `CPERT` and from below by `PARA(1)`.
DPERT	Matrix of maximum perturbation values for the direct transmittance matrix (`D`). The perturbation value for each element of `D` will be bounded from above by the corresponding element of `DPERT` and from below by `PARA(1)`.
TS	Sample time for `dlinmod`. Defaults to the maximum sample time for the model.

The steps for using the linearization tools are as follows:

1 Prepare a SIMULINK model configured with Inport blocks for the inputs, and Outport blocks for the outputs. Note that blocks from the Sources block library do not count as inputs, and blocks from the Sinks block library do not count as outputs. It is not necessary to have an Inport block, but there should be at least one Outport block.

2 Use the procedures discussed in Section 7.2 to determine the number and ordering of the model's states.

3 Execute the appropriate linearization command from the MATLAB command line or from within an M-file. If you are only interested in the system matrix, use a single output argument, such as

```
A = linmod('sysmdl_a')
```

If you need the entire linear state-space model, use all of the output arguments.

The Derivative block and Transport Delay block should not be used in models to be linearized, as they can cause numerical trouble for the linearization functions. Replace Derivative blocks with Switched Derivative blocks. The Switched Derivative block is in the Linearization block library, which is in turn found in the SIMULINK Extras library located in the Blocksets and Toolboxes block library.

Once you have obtained the linear model, you can use a variety of MATLAB commands to analyze the system dynamics or to design linear controllers.

Example 7-8

Consider the nonlinear system discussed in Example 7-6:

$$\begin{aligned} \dot{x}_1 &= -x_2 \\ \dot{x}_2 &= x_1 + x_2^3 - 3x_2. \end{aligned}$$

We wish to linearize this system about the origin. Applying Equation (7-7), we compute

$$A = \begin{bmatrix} 0 & -1 \\ 1 & -3 \end{bmatrix}.$$

A SIMULINK model of this system is shown in Figure 7-7. From the state portrait (Figure 7-12), it appears that this system exhibits stable behavior without oscillation. Thus we would expect both eigenvalues of A to be real and negative. Figure 7-13 shows how to verify this using MATLAB. First the system matrix is determined using `linmod` and then `eig` is used to compute the eigenvalues of the system matrix, which are indeed real and negative.

```
EDU» A = linmod('sysmdl_e')
A =
         0   -1.0000
    1.0000   -3.0000
EDU» eig(A)
ans =
   -0.3820
   -2.6180
```

Figure 7-13 Using `linmod` to compute the system (A) matrix of a nonlinear second-order system

Example 7-9

In this example we will discuss the design of a linear controller for a nonlinear system. Consider the cart with an inverted pendulum illustrated in Figure 7-14. This system approximates the dynamics of a rocket immediately after lift-off. The objective of the rocket control problem is to maintain the rocket in a vertical attitude while it accelerates. The objective in the control of this model is to move the cart to a specified position (x) while maintaining the pendulum vertical. Ogata [3] presents a detailed explanation of the design of a linear controller for this system using the technique of pole placement and MATLAB.

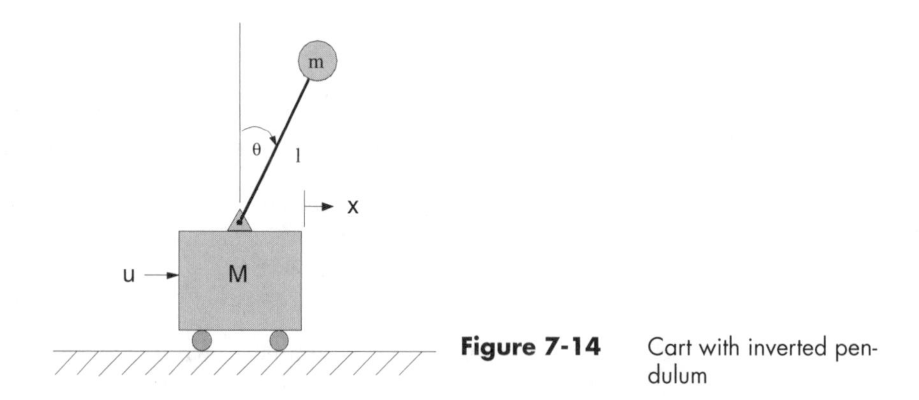

Figure 7-14 Cart with inverted pendulum

Referring to Figure 7-14, the input to the system is a horizontal force (u) applied to the cart of mass M. The pendulum is free to rotate without friction in the plane of the page. The pendulum is of length l, and the pendulum mass (m) is assumed to be concentrated at the end. The equations of motion of this system in terms of the cart displacement and pendulum angle can be written

$$(M+m)\ddot{x} - ml\dot{\theta}^2\sin\theta + ml\ddot{\theta}\cos\theta = u$$

$$m\ddot{x}\cos\theta + ml\ddot{\theta} = mg\sin\theta\,,$$

where g is the acceleration due to gravity. Examining these equations, we notice that $\ddot{x}$ and $\ddot{\theta}$ appear in both equations. If we build a SIMULINK model using these equations, the model will contain an algebraic loop (see Chapter 8 for a detailed discussion of algebraic loops). Employing a little algebra, the equations of motion can be rewritten to eliminate the algebraic loop:

$$\ddot{x} = \frac{ml\dot{\theta}^2\sin\theta - mg\cos\theta\sin\theta + u}{M + m\sin^2\theta}$$

$$\ddot{\theta} = \frac{-(ml\dot{\theta}^2\cos\theta\sin\theta - mg\sin\theta + u\cos\theta - Mg\sin\theta)}{l(M + m\sin\theta^2)}.$$

To follow the procedure of Ogata [3], our next task is to build a SIMULINK model of the system, with u as the input and x as the output. We choose as state variables x, $\dot{x}$, θ, $\dot{\theta}$, and will need all four state variables in our controller. The model is enclosed in a subsystem with a scalar Inport block and a vector Outport block, as shown in Figure 7-15. Note that the equations of motion are computed using Function blocks, and the system parameters are variables defined in the MATLAB workspace. The subsystem is placed in a SIMULINK model that has a single input (u) and a single output (x), as shown in Figure 7-16. A Demultiplexer block decomposes the state vector into its components. x is routed to an Outport block, and the remaining components of the state vector are routed to Terminator blocks.

The system parameters are

$$M = 2 \text{ kg}$$
$$m = 0.1 \text{ kg}$$
$$l = 0.5 \text{ m}$$
$$g = 9.8 \text{ m/s}^2$$

The M-file shown in Figure 7-17 determines the model characteristics, producing the output shown in Figure 7-18.

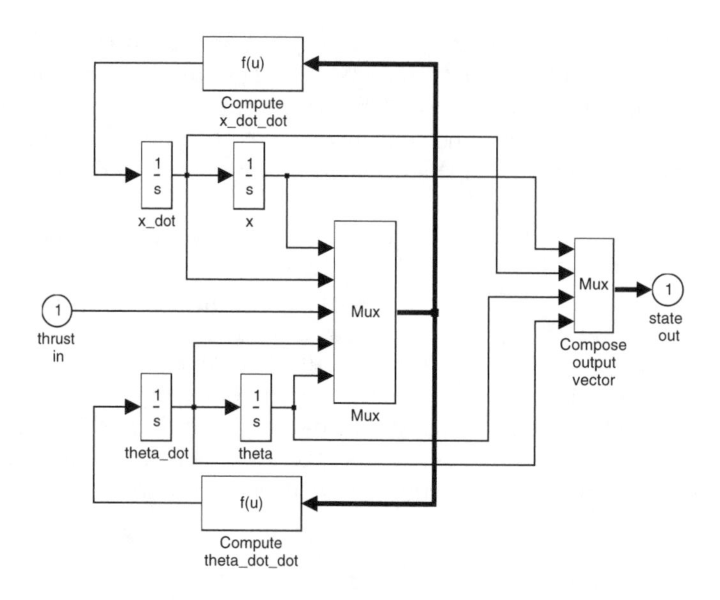

Figure 7-15 Inverted pendulum subsystem

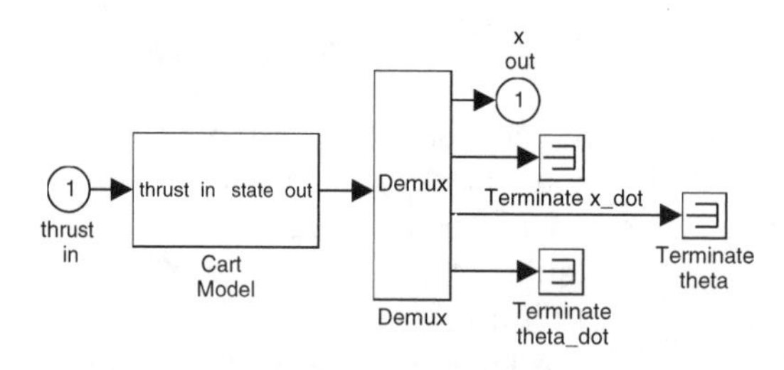

Figure 7-16 SIMULINK model to be linearized

```
% Initialize the system parameters and obtain the
% model characteristics
%
% System parameters
l = 0.5 ;   % Pendulum length
g = 9.8 ;   % Gravity
m = 0.1 ;   % Pendulum bob mass
M = 2 ;     % Cart mass
% Get the model characteristics
[sizes,x0,states] = sysmdl_h([],[],[],0)
```

Figure 7-17 M-file to determine the characteristics of the inverted pendulum model

```
EDU» crt_init
sizes =
      4
      0
      1
      1
      0
      1
      1
x0 =
      0
      0
      0
      0
states =
     'sysmdl_h/Cart  Model/x'
     'sysmdl_h/Cart  Model/x_dot'
     'sysmdl_h/Cart  Model/theta_dot'
     'sysmdl_h/Cart  Model/theta'
```

Figure 7-18 Inverted pendulum model characteristics

Note that the SIMULINK state vector is

$$\begin{bmatrix} x \\ \dot{x} \\ \dot{\theta} \\ \theta \end{bmatrix}$$

which is ordered differently from the state vector at the Outport block of the subsystem.

Next we incorporate the subsystem in a SIMULINK model that includes a state feedback controller with an integrator in the forward path, shown in Figure 7-19. The input to the control system is the desired cart position. The Step function block is configured to produce a unit step at 1.0 second. The output is the current cart position, displayed using a Scope block.

Figure 7-20 shows an M-file that computes the controller gains according to the procedure of Ogata [3]. The M-file uses `linmod` to obtain the state-space matrixes. Then Ackermann's method is used to compute the gains. The results of executing this M-file are shown in Figure 7-21. Finally, we test the controller using the M-file shown in Figure 7-22. This M-file runs the simulation and then plots the x and θ trajectories. The plots are shown in Figure 7-23.

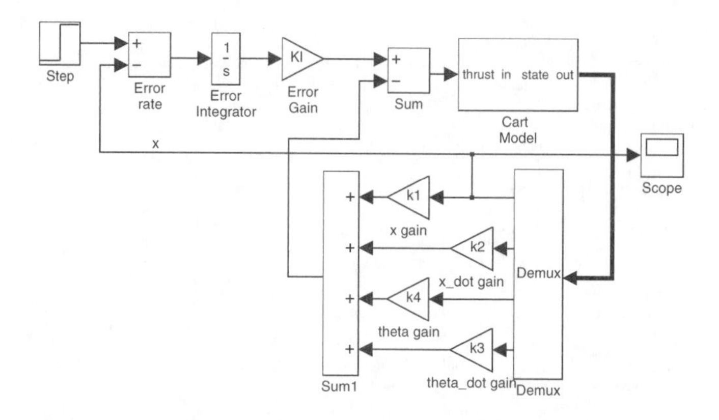

Figure 7-19 Inverted pendulum control SIMULINK model

```
% Design the feedback gain matrix for the cart with inverted pendulum
% example. This example follows the procedure outlined in Example 2-3 of
% "Designing Linear Control Systems with MATLAB", by K. Ogata
%
% Note that for this model, the state vector is
% [x ; x_dot ; theta_dot ; theta]', so the arrangement of the system
% matrixes will be different than in the Ogata example.
%
% System parameters
l = 0.5 ;   % Pendulum length
g = 9.8 ;   % Gravity
m = 0.1 ;   % Pendulum bob mass
M = 2 ;     % Cart mass
% Get the system matrixes
[A,B,C,D] = linmod('sysmdl_h',[0,0,0,0],[0])
A1 = [A zeros(4,1) ; -C 0] ;
B1 = [B ; 0] ;
% Define the controllability matrix M
MM = [B1 A1*B1 A1^2*B1 A1^3*B1 A1^4*B1] ;
% Check the rank of MM, must be 5 if system is completely controllable
rank(MM)
% Obtain the characteristic polynomial
J = [-1+sqrt(3)*i,0,0,0,0;0,-1-sqrt(3)*i,0,0,0;...
     0,0,-5,0,0;0,0,0,-5,0;0,0,0,0,-5] ;
Phi = polyvalm(poly(J),A1) ;
% Get the feedback matrix using Ackermann's formula
KK = [0,0,0,0,1]*(inv(MM))*Phi
k1 = KK(1)  ; % x gain
k2 = KK(2)  ; % x_dot gain
k3 = KK(3)  ; % theta_dot gain
k4 = KK(4)  ; % theta gain
KI = -KK(5) ; % error integrator gain
```

Figure 7-20 Controller design M-file

```
A =
         0    1.0000         0         0
         0         0         0   -0.4900
         0         0         0   20.5800
         0         0    1.0000         0
B =
         0
    0.5000
   -1.0000
         0
C =
     1     0     0     0
D =
     0
ans =
     5
KK =
  -56.1224  -36.7868  -35.3934 -157.6412   51.0204
```

Figure 7-21 Inverted pendulum linear model and controller gains

```
% Run the inverted pendulum simulation
opts = simset('InitialState',[0,0,0,0,0],'Solver','ode45') ;
[t,x] = sim('sysmdl_i',[0:0.1:10],opts) ;
% Plot the position and pendulum angle
subplot(2,1,1);
plot(t,x(:,1));
grid;
ylabel('x');
subplot(2,1,2);
plot(t,x(:,4));
grid;
xlabel('Time (sec)') ;
ylabel('theta');
```

Figure 7-22 M-file to test the inverted pendulum controller

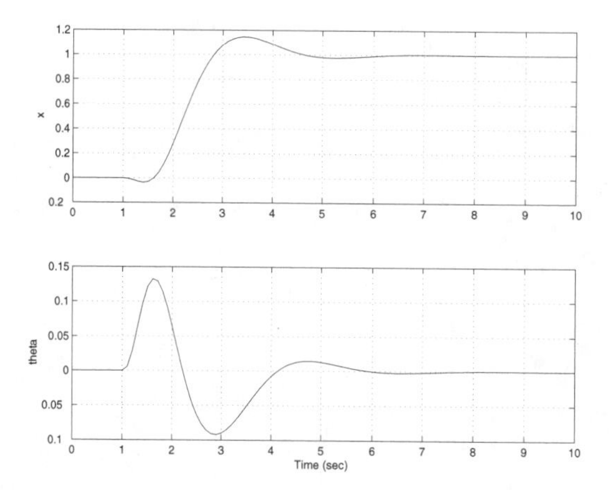

Figure 7-23 Inverted pendulum simulation results

7.5 Trim Tools

It is frequently useful to find equilibrium points of systems. As discussed in Example 7-6, an equilibrium point is a point at which all state derivatives are zero. For example, a valuable technique in nonlinear systems analysis is to linearize a system about an equilibrium point in order to assess the system stability in the vicinity of the equilibrium point using linear systems approximations. Another situation in which equilibrium point location is useful is the assessment of the accuracy of a control system, as the steady-state behavior of the system can be thought of as an equilibrium.

The SIMULINK `trim` command locates equilibrium points and also locates partial equilibrium points, where we define a partial equilibrium point to be a point at which selected state derivatives are zero, but other state derivatives are allowed to be nonzero. `trim` locates an equilibrium point by numerically searching for a point (values of `x`, `u`, `y`) such that the maximum absolute value of the state derivative is minimized. If elements of the state (`x`), input (`u`), or output (`y`) are fixed, the fixed elements are treated as constraints that the algorithm will attempt to satisfy. In this case the algorithm minimizes both the state derivative and the constraint error. The syntax of the `trim` command is

```
[x,u,y,dx] = trim(model,x0,u0,y0,ix,iu,iy,dx0,idx,options,t)
```

All of the output arguments and all of the input arguments except *`model`* are optional. The arguments are defined in Table 7-5.

It is generally best not to overconstrain the problem. Thus if an output is the same as an element of the state vector, and that element is to be fixed, specify (using `x0`, `ix` or `y0`, `iy`) either the state vector element or the output vector element, but not both.

Table 7-5 `trim` command arguments

Argument	Meaning
`x`	Value of the state vector at the equilibrium point. If no output arguments are specified, `trim` will return `x`.
`u`	Value of the input vector at the equilibrium point. `u` will be the empty vector if the system has no inport blocks. Source blocks are not considered inputs to the system.
`y`	Value of the output vector at the equilibrium point. If there are no Outport blocks, `y` will be the empty vector.
`dx`	Value of the state derivative vector at the equilibrium point.
`model`	Name of the SIMULINK model, enclosed in single quotes ('), and without the file extension. For example, if the model is stored in the file `s_examp.mdl`, the first input argument is `'s_examp'`. This input argument is required.
`x0`	Initial guess for the state vector. The algorithm will begin the search for the equilibrium at this point. Some elements of the state vector may be fixed at the value specified in `x0`, using argument `ix`. `x0` must be a column matrix, for example `[0;1;0;0]`. If `x0` is specified, all elements must be present.
`u0`	Initial guess for the input vector. Some elements of the input vector may be fixed at the value specified in `u0`, using argument `iu`.
`y0`	Initial guess for the output vector. Some elements of the output vector may be fixed at the value specified in `y0`, using argument `iy`.
`ix`	Vector indicating which elements of the state vector are to be fixed (that is, treated as constraints). For example, if the second and fourth elements of the state vector are to be fixed at the value specified in `x0`, `ix` would be `[2,4]`.
`iu`	Vector indicating which elements of the input vector are to be fixed (that is, treated as constraints). For example, if the second and fourth elements of `u0` are to be treated as constraints, `iu` would be `[2,4]`.

Table 7-5 `trim` command arguments (Continued)

Argument	Meaning
`iy`	Vector indicating which elements of the output vector are to be fixed (that is, treated as constraints). For example, if the second and fourth elements of `y0` are to be treated as constraints, `iy` would be `[2,4]`.
`dx0`	State derivative vector at a partial equilibrium point. `dx0` is used in conjunction with `idx` to fix certain elements of the state derivative.
`idx`	Vector indicating which elements of the state derivative vector are to be fixed at the value specified in `dx0`. For example, if the third element of the state derivative is to be fixed at `dx0(3)`, `idx` would be `[3]`. The remaining elements of the state derivative vector will be free.
`options`	Vector of optimization options that will be passed to the constrained optimization function. For details on `options`, refer to the MATLAB help screen for `FOPTIONS`. This argument is usually omitted or left empty (`[ ]`). If `trim` fails to converge, changing optimizations parameters may help.
`t`	Value of time at which to locate the equilibrium. This input argument needs to be specified only if the state derivative is time dependent.

Input arguments `x0`, `u0`, and `y0` define the starting point for the search. There is no guarantee, however, that `trim` will locate the equilibrium closest to the starting point. Additionally, `trim` may not converge, even if an equilibrium point exists. If `trim` fails to converge, it frequently helps to try a different starting point.

Example 7-10

Consider the differential system

$$\dot{x}_1 = x_1^2 + x_2^2 - 4$$
$$\dot{x}_2 = 2x_1 - x_2$$

We wish to locate the equilibrium points of this system and assess the system stability at each equilibrium point. A SIMULINK model of this system is shown in Figure 7-24. Figure 7-25 shows a state portrait for this system, produced using a suitably modified version of the M-file shown in Figure 7-11. We note that there appear to be two equilibrium points, located near (–1, –2) and (1, 2). Furthermore, from the state portrait, it appears that the equilibrium near (–1, –2) is stable, and the equilibrium near (1, 2) is unstable.

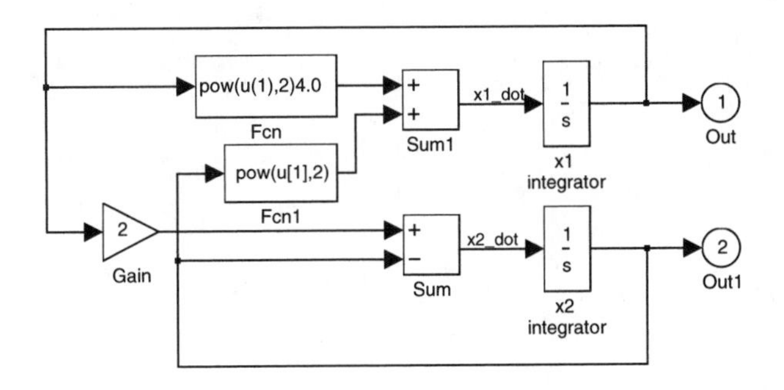

Figure 7-24 SIMULINK model of nonlinear system

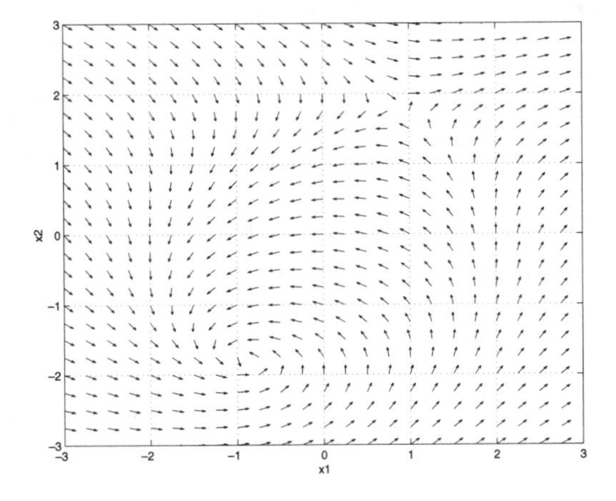

Figure 7-25 State portrait

To assess the stability of the equilibrium point, we first locate the equilibrium points using `trim`. Next, using `linmod`, we linearize the system about each equilibrium point. Finally, we use MATLAB function `eig` to compute the eigenvalues of the linearized system at each equilibrium point. An M-file that performs these operations is shown in Figure 7-26, and the results of executing this M-file are shown in Figure 7-27.

Examining the results in Figure 7-27, we can see that the equilibrium point near (–1, –2) is actually located at (–0.89, –1.79). Since the real parts of both eigenvalues at this point are negative, we can conclude that the system is stable in the vicinity of this equilibrium point (this equilibrium point is called

```
% Locate the two equilibrium points for the system
% Linearize the system about each equilibrium,
% then compute the eigenvalues at each equilibrium.
xa = trim('sysmdl_1',[-1;-2]) ;  % Locate the equilibrium
A = linmod('sysmdl_1',xa) ;
eigv_a = eig(A) ;
fprintf('Eigenvalues at (%f,%f):\n',xa) ;
eigv_a
xb = trim('sysmdl_1',[1;2]) ;  % Locate the other equilibrium
A = linmod('sysmdl_1',xb) ;
eigv_b = eig(A) ;
fprintf('Eigenvalues at (%f,%f):\n',xb) ;
eigv_b
```

Figure 7-26 M-file to assess system stability using linearization at the equilibrium points

```
Eigenvalues at (-0.894427,-1.788854):
eigv_a =
  -1.3944+ 2.6457i
  -1.3944- 2.6457i
Eigenvalues at (0.894427,1.788854):
eigv_b =
    3.4110
   -2.6222
```

Figure 7-27 Equilibrium assessment results

a *stable focus*). One of the eigenvalues at the equilibrium point at (0.89,1.79) is negative, and the other is positive. Thus the system behavior in the vicinity of this point should be unstable (this type of equilibrium is called a *saddle point*).

Example 7-11

Consider the rotating pendulum shown in Figure 7-28. The system consists of a motor driving a pendulum of length l, hinged at O. The mass (m) of the pendulum is assumed to be concentrated at the end. The rotating parts of the motor are modeled as a flywheel with moment of inertia I_m. The input to the system is the motor torque (τ). Friction and aerodynamic drag may be ignored. Our task is to determine the motor angular velocity ($\dot{\phi}$) that is consistent with a constant pendulum deflection (θ) of 0.5 rad.

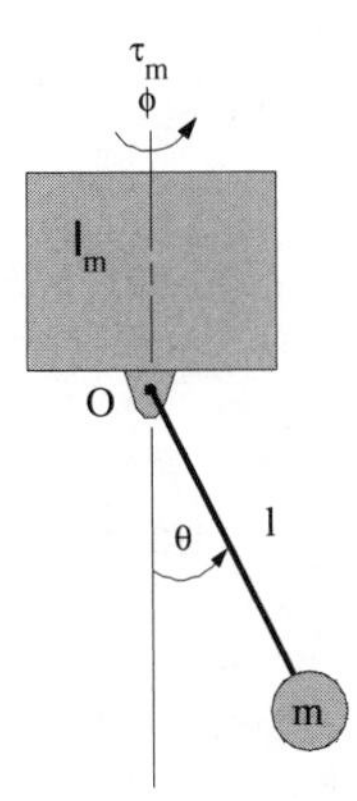

Figure 7-28 Rotating pendulum

We start by writing the equations of motion in terms of the shaft angle (ϕ) and the pendulum deflection (θ):

$$\ddot{\theta} = \frac{(l\dot{\phi}^2\cos\theta - g)\sin\theta}{l}$$

$$\ddot{\phi} = \frac{\tau - 2ml^2\dot{\phi}\dot{\theta}\sin\theta\cos\theta}{I_m + ml^2\sin^2\theta},$$

where g represents the acceleration due to gravity. Figure 7-29 shows a SIMULINK model of this system. The equations of motion are in the two function blocks.

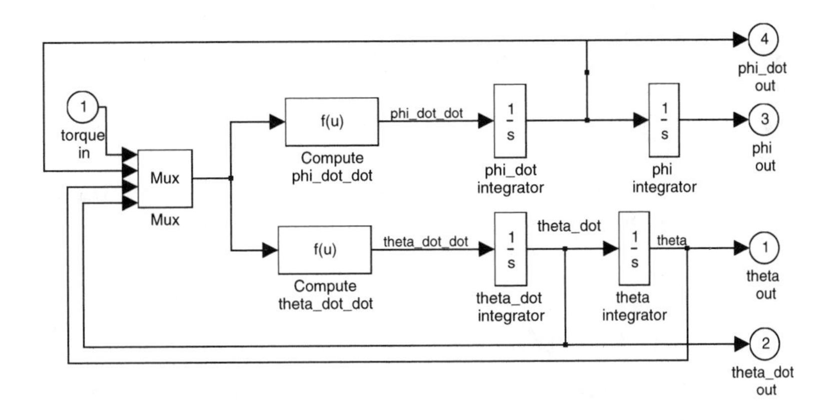

Figure 7-29 SIMULINK model of rotating pendulum

The MATLAB code in Figure 7-30 sets the model parameters and verifies that the ordering of the states is the same as the order of the model Outport blocks.

```
% Set the parameters for the rotating pendulum.
% Verify the ordering of the states.
% System parameters
l = 0.3 ;                % Pendulum length (meters)
I_m = 0.75 ;            % Motor moment of inertia (kilograms*meters^2)
g = 9.8 ;                % Gravity (meters/sec^2)
m = 0.2 ;                % Pendulum bob mass (kg)
[sizes,x0,states]=sysmdl_k([],[],[],0) ;
states
```

Figure 7-30 Set system parameters for rotating pendulum model

The pendulum deflection at the desired equilibrium is 0.5 rad, and the motor angular velocity is not zero, so we choose the initial state vector

```
x0=[0.5;0;0;5]
```

Since the system is conservative, we expect the input to be zero at equilibrium, so we set

```
u0 = 0
```

The output vector is the same as the state vector, so we set

```
y0 = [0.5;0;0;5]
```

We will allow the state and control to vary, but require that the output pendulum deflection be 0.5 rad. We set `ix` and `iu` to the empty vector and set

```
iy = 1
```

The value of the derivative of motor position (ϕ) (the third element of the state vector) will be nonzero at equilibrium, and all the other state derivatives will be zero, so we set

```
dx0 = [0;0;5;0]
idx = [1,2,4]
```

The MATLAB statements in Figure 7-31 execute the trim command with these values. The last element of `x` is the motor angular velocity at equilibrium, computed to be 6.1011 rad/s. You can easily verify that this value results in a balance between centrifugal force and gravity.

```
x0=[0.5;0;0;5] ;
u0 = 0 ;
y0 = [0.5;0;0;5] ;
ix = [] ;
iu = [] ;
iy = 1  ;
dx0 = [0;0;5;0] ;
idx = [1,2,4] ;
x = trim('sysmdl_k',x0,u0,y0,ix,iu,iy,dx0,idx)
x =
    0.5000
         0
    0.0000
    6.1011
```

Figure 7-31 Locating the rotating pendulum equilibrium point

7.6 Summary

In this chapter, we have discussed the use of SIMULINK analysis tools. First we showed how to identify the structure of a model's state vector. Next we showed how to run SIMULINK models from the MATLAB workspace using `sim`. Then we used the linearization tools to find linear models of nonlinear systems, and used the linear models for control system design. Finally, we discussed using `trim` to locate equilibrium points.

7.7 References

1 Khalil, Hassan K., *Nonlinear Systems*, 2nd ed. Upper Saddle River, N.J.: Prentice Hall, 1996. An excellent text on the analysis and design of nonlinear systems.

2 Ogata, Katsuhiko, *Modern Control Engineering*. Englewood Cliffs, N.J.: Prentice Hall, 1990. This book presents a thorough coverage of the standard techniques for the analysis and design of controls for continuous systems.

3 Ogata, Katsuhiko, *Designing Linear Control Systems with MATLAB*. Englewood Cliffs, N.J.: Prentice Hall, 1993, pp. 50–67. This book presents brief tutorials and MATLAB implementations of several important linear systems design techniques, including pole placement, state observers, and linear quadratic regulators.

4 Shahian, Bahram, and Hassul, Michael, *Control System Design Using MATLAB*. Englewood Cliffs, N.J.: Prentice Hall, 1993. This book provides an introduction to MATLAB programming and uses MATLAB to solve many of the standard problems in classical control and modern control theory.

5 Scheinerman, Edward C., *Invitation to Dynamical Systems*. Upper Saddle River, N.J.: Prentice Hall, 1996, p. 134. This text provides a good introduction to the analysis of nonlinear systems.

6 Vidyasagar, M., *Nonlinear Systems Analysis*, 2nd ed. Englewood Cliffs, N.J.: Prentice Hall, 1993. This book presents a detailed coverage of the analysis of nonlinear systems.

8

Numerical Issues

In this chapter, we will discuss two numerical issues you should consider when building SIMULINK models. First, we will consider choosing the best differential equation solver for a particular model, which can frequently improve the speed and accuracy of a simulation. Then we will explain algebraic loops, how SIMULINK deals with them, and how you can eliminate them if necessary.

8.1 Introduction

The SIMULINK block diagram metaphor for programming frees you from many of the details of writing a computer program to model a dynamical system. However, when you build a SIMULINK model, you are programming and, as with all programming tasks, there are numerical issues you should keep in mind. We will discuss two of these issues in this chapter: choosing a differential equation solver and dealing with algebraic loops.

8.2 Choosing a Solver

SIMULINK provides several differential equation solvers. The majority of the solvers are the result of recent numerical integration research and are among the fastest and most accurate methods available. Detailed descriptions of the algorithms are available in the paper by Shampine [4], available from The MathWorks.

It is generally best to use the variable-step solvers, as they continuously adjust the integration step size to maximize efficiency while maintaining a specified accuracy. The SIMULINK variable-step solvers can completely decouple the integration step size and the interval between output points, so it is not necessary to limit the step size to get a smooth plot or to produce an output trajectory with a predetermined fixed step size. The available solvers are listed in Table 3-2, repeated here for convenience in Table 8-1.

Solver Selection Considerations

There is no universal "best" differential equation solver. Choosing the best solver for a particular system requires understanding the system dynamics.

Table 8-1 SIMULINK Solvers

Solver Type	Characteristics
ODE45	Excellent general purpose single-step solver. Based on the Dormand-Prince fourth/fifth-order Runge-Kutta pair. ODE45 is the default solver and is usually a good first choice.
ODE23	Uses the Bogacki-Shampine second/third-order Runge-Kutta pair. Sometimes works better than ODE45 in the presence of mild stiffness. Generally requires a smaller step size than ODE45 to get the same accuracy.
ODE113	Variable-order Adams-Bashforth-Moulton solver. Since ODE113 uses the solutions at several previous time points to compute the solution at the current time point, it may produce the same accuracy as ODE45 or ODE23 with fewer derivative evaluations, and thus perform much faster. Not suitable for systems with discontinuities. See Kincaid and Cheney [2] for a good explanation of Adams-Bashforth-Moulton solvers.
ODE15S	Variable-order multistep solver for stiff systems. Based on recent research using numerical difference formulas. If a simulation runs extremely slowly using ODE45, try ODE15S.
ODE23S	Fixed-order single-step solver for stiff systems. Because ODE23S is a single-step method, it is sometimes faster than ODE15S. If a system appears to be stiff, it is a good idea to try both stiff solvers to determine which performs the best.
Fixed- and Variable-Step Discrete	Special solvers for systems that have no continuous states.
ODE5	Fixed-step-size version of ODE45.
ODE4	Classic fourth-order Runge-Kutta formulas using a fixed step size.
ODE3	Fixed-step version of ODE23.
ODE2	Fixed-step-size second-order Runge-Kutta method, also known as Heun's method.
ODE1	Euler's method using a fixed step size.

Let's look briefly at the characteristics of each of the methods and mention systems for which each is probably the best choice.

ODE45 and ODE23 are Runge-Kutta methods. These methods approximate the solution function (in the case of SIMULINK, the solution function is the state trajectory of the SIMULINK model) by numerically approximating a Taylor series of a fixed number of terms, the *order* being defined as the highest derivative in the series. The principal error in a Taylor series approximation of a function is known as *truncation error*, and is due to the truncation of the Taylor series to a finite number of terms. ODE45 and ODE23 estimate the truncation error by computing the value of the state variables at the end of an integration step using two Taylor series approximations of different orders (4 and 5 or 2 and 3). The difference in the two computed values is a reliable indicator of the total truncation error. If the error is too large, the integration step size is reduced, and the integration step is repeated. If the error is too small (more accuracy than needed), the step size is increased for the next integration step. An important characteristic of algorithms such as ODE45 and ODE23 is that they select the intermediate points in the integration step such that both Taylor series approximations use the same derivative function evaluations.

ODE113 is a variable-order Adams method (namely, an Adams-Bashforth-Moulton method), a multistep predictor-corrector algorithm. The predictor step approximates the derivative function as a polynomial of degree $n-1$, where n is the order of the method. The coefficients of the predictor polynomial are computed using the previous $n-1$ solution points and the derivatives at the points. A trial next solution point is computed by extrapolation. Next a corrector polynomial is fit through the previous n points and the newly computed trial solution point, and this polynomial is evaluated to recompute the trial solution point. The corrector portion of the algorithm can be repeated to refine the solution point. The difference between the predictor solution and the corrector solution is a measure of the integration error and is used to adjust the integration step size. ODE113 also adjusts the degree of the approximating polynomials to balance accuracy and efficiency. Multistep methods such as ODE113 tend to work very well for systems that are smooth. They don't work well for systems with discontinuities because the polynomial approximation assumes a smooth function.

ODE15S is a variable-order multistep algorithm specifically designed to work well with stiff systems. A stiff system is one that has both very fast dynamics and very slow dynamics (widely separated eigenvalues). An example would be a system that has a sharp start-up transient followed by a relatively slow steady-state response. Special techniques, such as ODE15S, are required to model such systems accurately. These algorithms contain extra logic to detect transitions in a system's dynamics. The extra computational work expended in adapting to rapidly changing dynamics makes the stiff solvers inefficient for systems that are not stiff.

ODE23S is a single-step stiff system solver based on the Rosenbrock formulas. This method is of fixed order, and since there is no order adjustment logic, it is sometimes faster than ODE15S.

The discrete solver is a special method that is applicable only to systems that have no continuous states. Although all of SIMULINK's solvers are suitable for such systems, the discrete solver is the fastest choice for these systems.

The Solver options section contains four fields to control integration step size adjustment for the variable-step-size integrators. Two fields, **Max step size** and **Initial step size**, permit you to reduce the likelihood of the solver missing important system behavior. To allow SIMULINK to use its default values for these parameters, enter `auto` in the respective fields. The other two fields allow you to set the absolute and relative tolerances used in the step size adjustment logic.

The default **Max step size** is

$$h_{\text{max}} = \frac{t_{\text{stop}} - t_{\text{start}}}{50} \tag{8-1}$$

which is generally satisfactory. There may be certain situations in which it is desirable to enter a fixed value for **Max step size**. For example, if the duration of the simulation is extremely long, the default maximum step size may be too large to guarantee that no important behavior will be missed. If the system is known to be periodic, the performance of the step size adjustment logic may be slightly improved by limiting the maximum step size to a fraction (The MathWorks suggests ¼) of the period. It is not advisable to set the maximum step size to limit the spacing between points in the output trajectory, as it is much more economical computationally to use the Output options section of the solver page for that purpose.

If **Initial step size** is set to `auto`, SIMULINK will compute the initial step size based on the state derivatives at the start of the simulation. If the system dynamics are believed to contain a sharp transient soon after t_{start}, set **Initial step size** to a value small enough to permit the solver to detect the transient. In most other situations, it is best to allow SIMULINK's built-in logic to compute the initial step size.

Relative tolerance and **Absolute tolerance** are used to compute the allowable value of integration error estimate (e_i) for each state (x_i) according to the formula

$$e_i \le \max(tol_{Rel}|x_i|, tol_{Abs}) \tag{8-2}$$

where tol_{Rel} and tol_{Abs} are **Relative tolerance** and **Absolute tolerance**, respectively. The two tolerances for a hypothetical state are depicted graphically in Figure 8-1. In the region in which the magnitude of the state is large, tol_{Rel} determines the error bound. In the region where the magnitude of the state is small, tol_{Abs} determines the error bound. If the integration error estimate for any state exceeds its limit, the integration step size is reduced. If the error limits for every state exceed the estimates by some value that depends on the particular solver used, the integration step size for the subsequent step is increased.

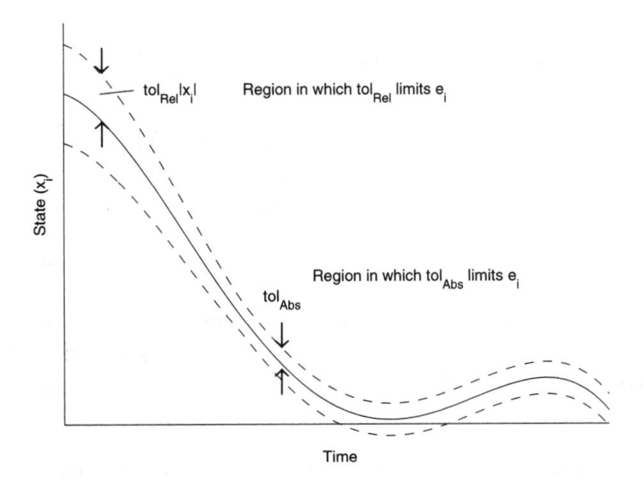

Figure 8-1 Solver error tolerance computation

The fixed-step solvers are all fixed-order single-step methods. These methods might be preferable if the system dynamics are sufficiently well understood that a nearly optimal integration step size is known. In such a situation, the elimination of step size and order adjustment might significantly speed up the simulation. It is not advisable to use a fixed-step-size integrator to force SIMULINK to produce an output trajectory with a fixed spacing between points, as the Output options section of the Solver page allows you to do that much more economically. If a fixed-step-size integrator is chosen, there is a single field, **Fixed step size**, to enter the step size. This field may contain a value for step size, or may be set to `auto` to allow SIMULINK to choose the fixed step size automatically.

Example 8-1

A stiff system is a system that has both fast dynamics and slow dynamics. Typically, the primary interest is in the slow dynamics, but ignoring the fast dynamics can cause the simulation to produce incorrect results. Consider the unforced second-order system

$$\ddot{x} + 100\dot{x} + 0.9999x = 0$$

Assume that the system starts at rest with $x = 1$. Taking the Laplace transform,

$$s^2X(s) - sx(0) - \dot{x}(0) + 100(sX(s) - x(0)) + 0.9999X(s) = 0$$

Substituting $x(0) = 1$, $\dot{x}(0) = 0$, and solving for $X(s)$,

$$X(s) = \frac{s + 100}{s^2 + 100s + 0.9999}$$

Taking the inverse Laplace transform, we get the time response:

$$x(t) = 0.0001e^{-99.99t} + 1.0001e^{-0.01t}$$

The response of this system has two components. The first component starts at a very small magnitude (0.0001) and decays rapidly. The second component starts at a magnitude 10,000 times as large and decays 10,000 times as slowly. So the slow response dominates the behavior of the system.

Now let's see what happens when we model this system. Figure 8-2 shows a SIMULINK model of the system. Set the **Initial condition** of the velocity Integrator to `0` and **Initial condition** of the displacement Integrator to `1`. Set **Start time** to `0` and **Stop time** to `500`. Select (check) **Simulation:Parameters** dialog box Workspace I/O page field **Save to workspace**, **Time**, and leave **States** and **Output** unselected.

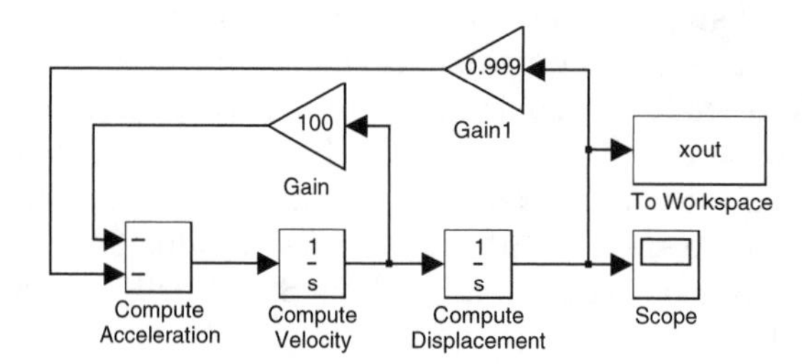

Figure 8-2 SIMULINK model of stiff second-order system

Next let's experiment with the different solvers to see the effect of the stiffness. First run the simulation using ODE15S. The simulation runs in a few seconds. The `tout` (produced as a result of selecting **Save to workspace, Time**) and `xout` (produced by the To Workspace block) vectors sent to the MATLAB workspace each have about 100 elements. Figure 8-3 shows the simulation results plotted using the MATLAB `plot` command.

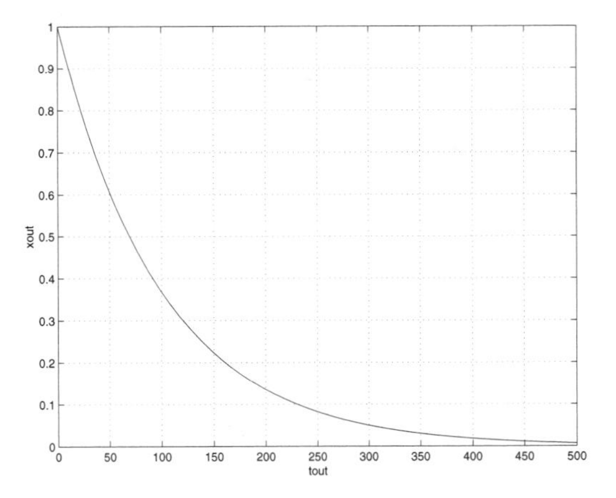

Figure 8-3 Stiff system results using ODE15S

Running the simulation using ODE45 takes a much longer time but produces a plot that appears identical to Figure 8-3. This time, however, the vectors `tout` and `xout` have approximately 15,000 elements each. The following MATLAB statements will produce a plot showing the change in the trajectory between successive points from point 14,000 to 14,100:

```
EDU» t=tout(14000:14100);
EDU» x=xout(14001:14101)-xout(14000:14100);
EDU» plot(t,x)
```

The plot appears in Figure 8-4. Inspecting the plot, you can see that there is a high-frequency component to the output trajectory resulting from the fast dynamics. This component is of very small magnitude (10^{-6}).

Finally, recall that we said that ignoring the fast dynamics can produce incorrect results. The stiff solver required about 100 time points, or on average 5 seconds between time points. Run the simulation using the fixed-step solver ODE4 and a 5-second step size. The simulation diverges, causing SIMULINK to issue an error message.

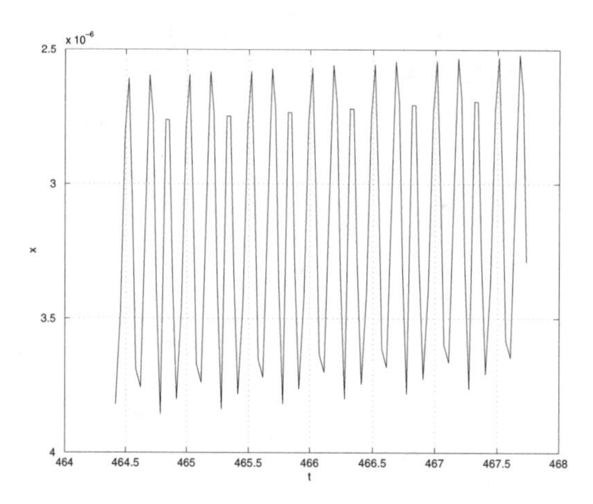

Figure 8-4 Oscillation of stiff system results using ODE45

8.3 Algebraic Loops

Algebraic loops are a programming issue that sometimes requires special care in SIMULINK modeling. An algebraic loop is a condition in which the output of a block drives the input of the same block. Consider the SIMULINK model in Figure 8-5. From the model we can compute

$$\dot{x} = u - k_3 k_2 \dot{x} - k_3 k_1 x\,. \tag{8-3}$$

So $\dot{x}$ is a function of x, u, and $\dot{x}$. Each integration step SIMULINK must solve the algebraic equation for $\dot{x}$. SIMULINK can't solve for $\dot{x}$ symbolically, so it uses an iterative numerical technique to solve the algebraic equation. This iterative procedure takes time, and in some cases SIMULINK fails to arrive at a solution.

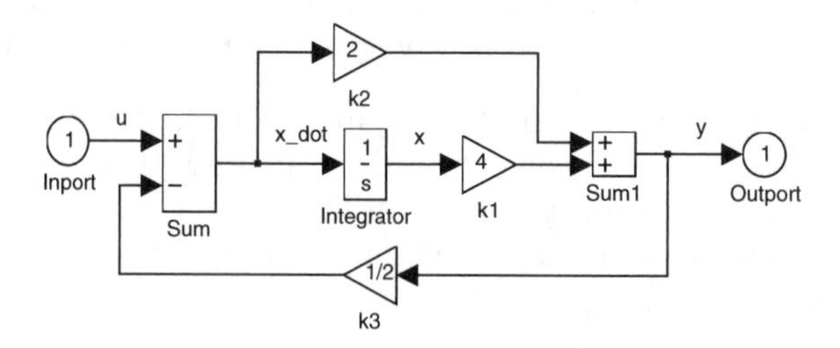

Figure 8-5 First-order system with an algebraic loop

In this simple example the algebraic loop is formed by Gain blocks `k2` and `k3` and Sum blocks `Sum` and `Sum1`. However, algebraic loops are not restricted to Gain and Sum blocks. Any block for which the current value of the output depends, even partially, on the current value of the input can be part of an algebraic loop. Such blocks are said to have *direct feedthrough*. Nearly all of the blocks in the Nonlinear block library have direct feedthrough. Transfer Fcn and Zero-Pole blocks for which the degree of the numerator is the same as the degree of the denominator and State-Space blocks with nonzero direct transmittance (`D`) matrixes also exhibit direct feedthrough.

Algebraic loops are not a problem unique to SIMULINK. If you attempt to use Equation (8-3) in a program in an algorithmic language such as FORTRAN, C, or even MATLAB, you will have to deal with the fact that $\dot{x}$ appears on both sides of the equation. While the scalar example here is easily dealt with, the situation is not always so simple. For example, modeling the dynamics of robot manipulators can produce complex nonlinear algebraic loops involving several state variables. The SIMULINK algebraic loop solver can frequently relieve you of the need to worry about algebraic loops, but you should understand the problem and how SIMULINK approaches it.

8.3.1 Newton-Raphson Method

SIMULINK attempts to solve algebraic loops using a robust implementation of the Newton-Raphson technique. To illustrate the basic idea of the technique, we will discuss the classic Newton-Raphson method for a scalar problem. For the preceding scalar example, the algebraic problem can be written

$$\dot{x} = f(x, \dot{x}, u, t), \tag{8-4}$$

where f is in general a nonlinear function. The Newton-Raphson method is an iterative process that attempts to solve the error function

$$\phi = \dot{x} - f(x, \dot{x}, u, t) = 0 \tag{8-5}$$

by minimizing the quadratic

$$P = \phi^2 \tag{8-6}$$

with respect to $\dot{x}$. At each iteration, compute (dropping the arguments of ϕ for convenience)

$$\Delta\dot{x} = -\phi/\frac{\partial\phi}{\partial\dot{x}}, \tag{8-7}$$

and then update the estimate

$$\dot{x}_{new} = \dot{x}_{old} + \Delta\dot{x} \tag{8-8}$$

The function ϕ and its partial derivative are evaluated at the current estimate of $\dot{x}$, and the current known values of x, u, and t. If $\partial\phi/\partial\dot{x}$ is constant—that is, if $f(x, \dot{x}, u, t)$ is linear with respect to $\dot{x}$—the Newton-Raphson procedure will converge in exactly one iteration. On the other hand, if $f(x, \dot{x}, u, t)$ is nonlinear with respect to $\dot{x}$, the procedure may require many iterations and may fail to converge.

8.3.2 Eliminating Algebraic Loops

SIMULINK will report the detection of an algebraic loop if the **Simulation:Parameters** Diagnostic choice for algebraic loops is set to **warning** or **error**. If an algebraic loop is detected, you have two options: leave the algebraic loop intact or eliminate it. If the speed of execution of the model is acceptable, leaving the loop intact is probably the better choice. If the speed of execution is not adequate, you must eliminate the algebraic loop.

The most desirable method of eliminating an algebraic loop is to reformulate the model into an equivalent model that does not have an algebraic loop. For example, the model shown in Figure 8-6 has the same input-output behavior as the model in Figure 8-5. However, this model does not have an algebraic loop and will therefore execute faster.

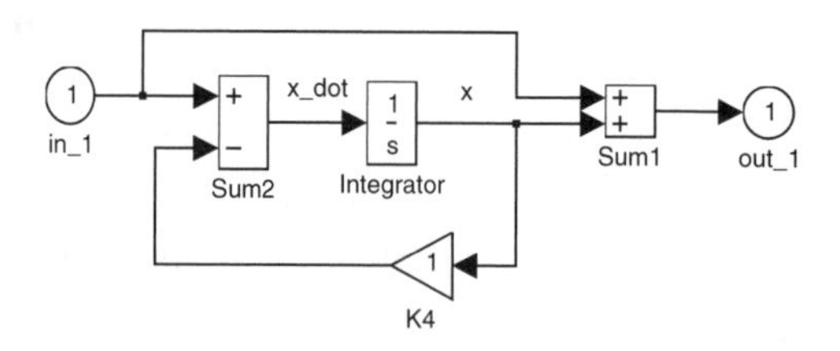

Figure 8-6 First-order system reformulated to eliminate the algebraic loop

It is not always convenient to reformulate a model such that there are no algebraic loops, as the equations of motion of some physical systems lead to algebraic loops. It is always possible in these situations to break an algebraic loop using a Memory block. Figure 8-7 shows the model of Figure 8-5 modified such that the algebraic loop is broken using a Memory block. While this approach can always be used to break an algebraic loop, it is not always satisfactory because the delay introduced by a Memory can degrade the accuracy of the simulation. As we will see in Example 8-2, this approach can change the behavior of the SIMULINK model so that it no longer accurately represents the physical system.

A final method of eliminating an algebraic loop is to use a MATLAB Fcn block to solve the algebraic problem directly. Although this approach requires that

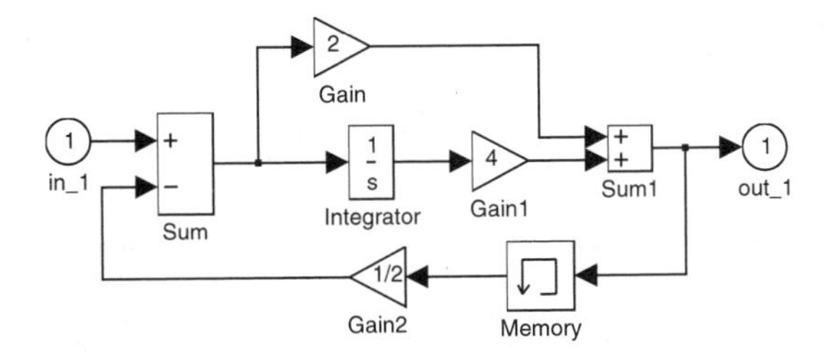

Figure 8-7 Breaking an algebraic loop with a memory block

you write MATLAB code to replace some SIMULINK blocks, it is sometimes the best solution.

Example 8-2

Consider the system of two carts shown in Figure 8-8. Cart 1 moves on the ground, and cart 2 moves relative to cart 1. x_1 and x_2 are the positions of the carts relative to inertial space. The angular displacements of the wheels are θ_1 and θ_2, and the total torque input to the wheels of cart 1 is τ_1, and to the wheels of cart 2 is τ_2.

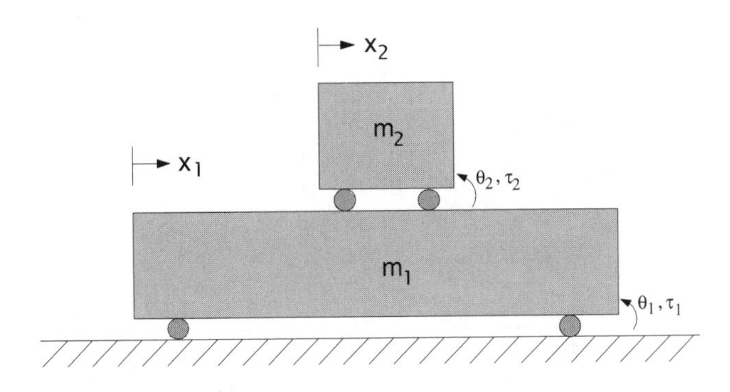

Figure 8-8 Directly coupled carts

The wheel angular displacements are related to the cart absolute positions by

$$x_1 = -r_1\theta_1$$
$$x_2 = x_1 - r_2\theta_2$$

The kinetic energy of the system in terms of the wheel angular velocities is

$$T = \frac{1}{2}m_1(-r_1\dot{\theta}_1)^2 + \frac{1}{2}m_2(-r_1\dot{\theta}_1 - r_2\dot{\theta}_2)^2$$

Using a Lagrangian approach, we solve for the wheel torques:

$$\begin{aligned}\tau_1 &= m_1 r_1^2\ddot{\theta}_1 + m_2 r_1^2\ddot{\theta}_1 + m_2 r_1 r_2\ddot{\theta}_2 \\ \tau_2 &= m_2 r_1 r_2\ddot{\theta}_1 + m_2 r_2^2\ddot{\theta}_2\end{aligned}$$

Solving for the angular accelerations, we get the coupled differential equations

$$\begin{aligned}\ddot{\theta}_1 &= K_{11}(\tau_1 - K_{21}\ddot{\theta}_2) \\ \ddot{\theta}_2 &= K_{22}(\tau_2 - K_{12}\ddot{\theta}_1)\end{aligned}$$

where

$$\begin{aligned}K_{11} &= \frac{1}{(m_1 + m_2)r_1^2} \\ K_{21} &= K_{12} = m_2 r_1 r_2 \\ K_{22} &= \frac{1}{m_2 r_2^2}\end{aligned}$$

A SIMULINK model that implements the equations of motion is shown in Figure 8-9. We set the parameters using the M-file shown in Figure 8-10.

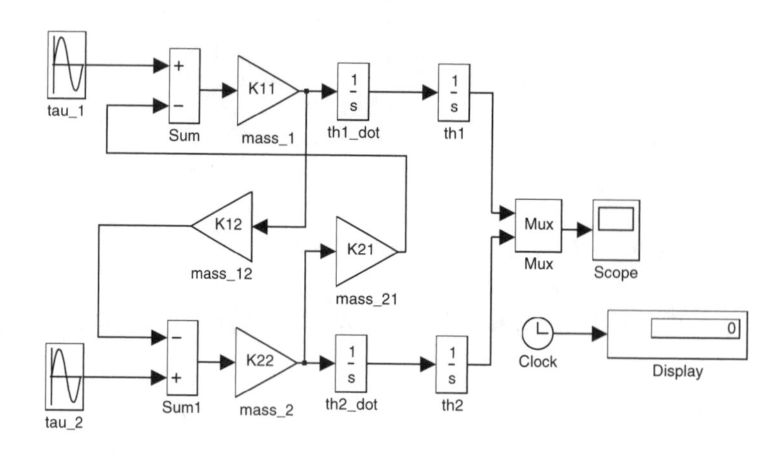

Figure 8-9 SIMULINK model of directly coupled carts

```
% Set up block parameters for algebraic loop example
m1 = 10 ;
m2 = 5 ;
r1 = 2 ;
r2 = 1.5 ;
K11 = 1/((m1+m2)*r1^2) ;
K12 = m2*r1*r2 ;
K21 = K12 ;
K22 = 1/(m2*r2^2) ;
```

Figure 8-10 MATLAB script to initialize coefficients

The input torques are

$$\begin{aligned} \tau_1 &= \sin(0.1t) \\ \tau_2 &= \sin(0.2t - 1) \end{aligned}$$

Executing the simulation produces an algebraic loop warning in the MATLAB workspace. The output trajectories of the two carts for the first 500 seconds are shown in Figure 8-11.

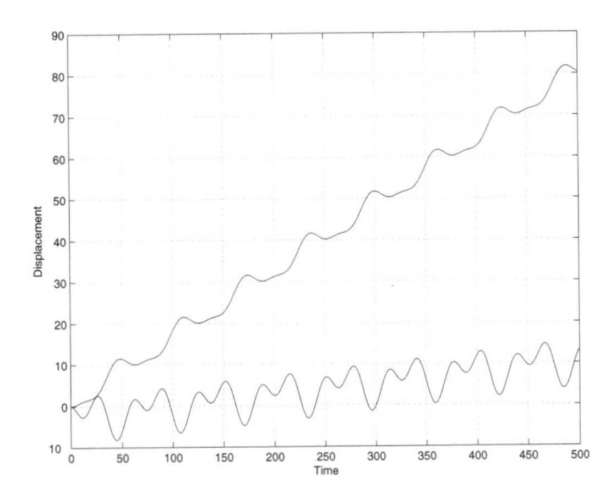

Figure 8-11 Directly coupled cart trajectories

To illustrate the potential for simulation errors when an algebraic loop is broken using a Memory block, consider the revised model in Figure 8-12. The Memory block eliminates the algebraic loop and causes the simulation to execute significantly faster. Unfortunately, the results of the simulation, shown in Figure 8-13, are incorrect. If the simulation is run with the **Max step size** on the **Simulation:Parameters** Solver page set to a sufficiently small value, correct results are produced. In this example, if the **Max step size** is set to `0.1`, the results are nearly identical to the results shown in Figure 8-11. Of course, limiting **Max step size** slows the simulation, negating the benefit of the Memory block.

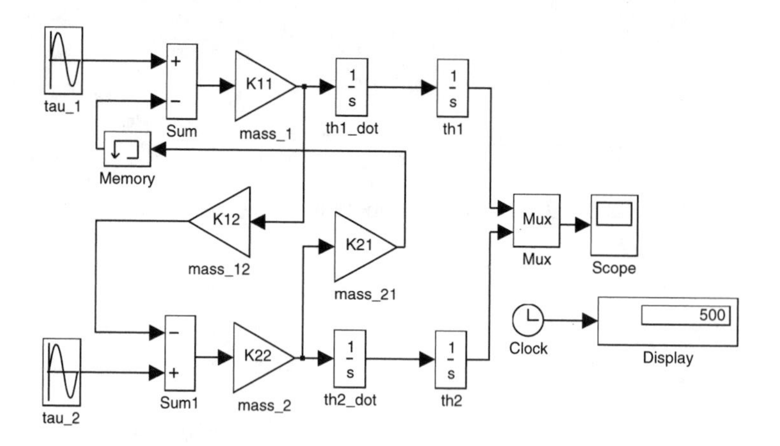

Figure 8-12 Directly coupled cart model with algebraic loop broken

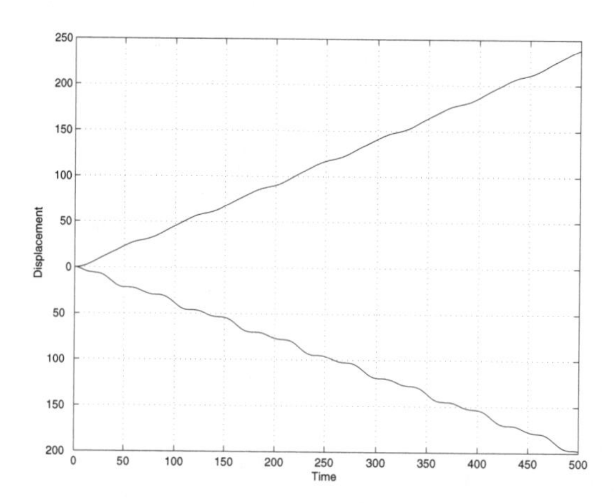

Figure 8-13 Trajectories using the directly coupled cart model with the algebraic loop broken

8.4 Summary

In this chapter we have discussed two important numerical issues you should consider when building SIMULINK models. First we discussed choosing the most appropriate solver, which can improve the speed and accuracy of a simulation. Then we discussed algebraic loops, how SIMULINK deals with them, and how you can eliminate them.

8.5 References

1 Hartley, Tom T., Beale, Guy O., and Chicatelli, *Digital Simulation of Dynamic Systems: A Control Theory Approach*. Englewood Cliffs, N.J.: Prentice Hall, 1994, pp. 190–238. In addition to covering the most important differential equation solution techniques, this book provides a detailed discussion of solutions to stiff systems.

2 Kincaid, David, and Cheney, Ward, *Numerical Analysis*. Pacific Grove, Calif.: Brooks/Cole Publishing Co., 1991, pp. 508–514. This is an excellent and very rigorous text covering many important topics in numerical analysis.

3 Mathews, John H., *Numerical Methods for Mathematics, Science, and Engineering*. Englewood Cliffs, N.J.: Prentice Hall, 1992. This is a fine numerical methods reference for scientists and engineers. In particular, Chapter 4, on interpolation and polynomial approximation, and Chapter 9, on differential equation solution, will assist you in understanding SIMULINK's solvers. Chapter 2 provides a rigorous treatment of Newton-Raphson methods. Many algorithms are provided, and MATLAB code for the algorithms is available from the MathWorks Web site (`www.mathworks.com`).

4 Shampine, Lawrence F., and Reichelt, Mark W., "The MATLAB ODE Suite," The MathWorks, Inc., Natick, Mass., 1996. This technical paper is available directly from The MathWorks. It provides a detailed discussion of the MATLAB differential equation solvers that are available from within SIMULINK. It also provides an extensive list of references.

Appendix: Block Reference

This appendix provides a brief listing of all the blocks in the SIMULINK 2 block libraries. For detailed descriptions of the blocks and their use, refer to the Block Browser.

SIMULINK Block Library

Block	Purpose
Sources	The Sources block library contains blocks that generate signals. The blocks include constants, sine waves, step functions, and a signal generator.
Sinks	The Sinks block library contains blocks that display signals, or save signals for further processing or analysis.
Discrete	The Discrete block library contains blocks that are used to model discrete systems. This library includes blocks that implement time delays, discrete integrators, discrete transfer functions, and zero- and first-order holds.
Linear	The Linear block library contains blocks used to model linear continuous systems. This library includes gain blocks (which can also be used with discrete systems), integrators, transfer function blocks, and a state-space block.
Nonlinear	The Nonlinear block library contains a variety of blocks used to model nonlinearities in both continuous and discrete systems. Examples of nonlinear blocks are logical and relational logic, nonlinear functions, sign and saturation functions, and lookup tables.
In Out Connections	The Connections block library contains a variety of blocks that transfer signals. Examples are goto blocks, input and output ports, and multiplexers and demultiplexers.

Sources Block Library

Block	Purpose
Constant	Generate a constant value. The constant can be a scalar or a vector.
Signal Generator	Generate a periodic signal (sine wave, square wave, or sawtooth wave) or random noise. Configuration parameters are signal amplitude and frequency.
Ramp	Generate a signal for which the time derivative is a constant.
Sine Wave	Generate a sine wave. Amplitude, phase, and frequency can be set.
Step	Generate a step function. Configuration parameters are step time, initial value, and final value.
Chirp Signal	Generate a sinusoidal signal of continuously increasing frequency.
Pulse Generator	Generate a rectangular wave. Configuration parameters are period, amplitude, duty cycle, and start time.
Repeating Sequence	Generate an arbitrary periodic signal. The signal is defined by a table of time points and amplitudes.
Clock	Generate a signal consisting of the current simulation time.
12:34 Digital Clock	Generate a signal consisting of the current simulation time, sampled at a specified period. Equivalent to a combination of a Clock block and a Zero-Order Hold, but much more efficient.

Sources Block Library (Continued)

Block	Purpose
untitled.mat From File	Generate a signal by interpolating in a MATLAB matrix stored in a file.
[T,U] From Workspace	Generate a signal by interpolating in a table defined by variables in the MATLAB workspace.
Random Number	Generate a signal containing normally distributed random numbers.
Band-Limited White Noise	Generate a signal containing band limited white noise of a specified power spectral density (PSD).

Sinks Block Library

Block	Purpose
Scope	Display scalar or vector signals in a method analogous to an oscilloscope.
XY Graph	Produce a graph using two scalar inputs. The signal connected to the top input port is the independent variable (x-axis), and the signal connected to the lower input port is the dependent variable (y-axis).
0 Display	Display the current value of the input signal.
untitled.mat To File	Save the input signal to a file in MATLAB .mat format. The signal may be scalar or vector.

Sinks Block Library (Continued)

Block	Purpose
simout To Workspace	Store the input signal in a MATLAB matrix accessible in the MATLAB workspace after the simulation stops. The signal may be scalar or vector.
STOP Stop Simulation	Cause the simulation to stop when the input signal is non-zero.

Discrete Block Library

Block	Purpose
1/z Unit Delay	The output signal of the Unit Delay is the input signal delayed by one sample time.
T/(z−1) Discrete–Time Integrator	The Discrete-Time Integrator is a discrete approximation to a continuous integrator.
Zero–Order Hold	The output of the Zero-Order Hold is the input at the most recent sample time.
First–Order Hold	The output of the First-Order Hold is a continuously varying signal. At an offset time δ since the last sample ($x(k)$), the output is $$x(k) + \frac{\delta}{T}(x(k) - x(k-1)),$$ where T is the sample period.
y(n)=Cx(n)+Du(n) x(n+1)=Ax(n)+Bu(n) Discrete State–Space	Model a linear time-invariant multiple-input, multiple-output discrete system using state-space notation.

Discrete Block Library (Continued)

Block	Purpose
$\frac{1}{1+2z^{-1}}$ Discrete Filter	The Discrete Filter block implements a discrete transfer function using the notation (polynomials of z^{-1}) frequently associated with digital filtering.
$\frac{1}{z+0.5}$ Discrete Transfer Fcn	The Discrete Transfer Fcn block implements a discrete transfer function using the notation (polynomials of z) frequently associated with control systems.
$\frac{(z-1)}{z(z-0.5)}$ Discrete Zero–Pole	Model a discrete transfer function using zero-pole notation.

Linear Block Library

Block	Purpose
1 Gain	The output of a Gain block is the input multiplied by a constant. The Gain block will work with scalar or vector signals, and the value of gain may be a scalar or vector compatible with the input signal.
+ + Sum	The output of the Sum block is the algebraic sum of its inputs. The number of inputs and the sign applied to each input can be set in the block dialog box.
$\frac{1}{s}$ Integrator	Compute the time integral of the input signal.
$\frac{1}{s+1}$ Transfer Fcn	Implement a continuous transfer function.
x' = Ax+Bu y = Cx+Du State–Space	Model a linear time-invariant multiple-input, multiple-output system, or subsystem using state-space notation.

Linear Block Library (Continued)

Block	Purpose
(s−1) / s(s+1) Zero–Pole	Implement a continuous transfer function using zero-pole notation.
du/dt Derivative	The output of the Derivative block is the time rate of change of the input.
• Dot Product	The Dot Product block accepts two vector signals of the same dimension. The output is the dot product of the current input vectors.
K Matrix Gain	The output of the Matrix Gain block is the current input vector multiplied by a compatible matrix. The input vector is treated as a column vector. Therefore, the gain matrix must have the same number of columns as there are elements in the input vector. The number of outputs is the same as the number of rows in the gain matrix, which does not have to be square.
1 Slider Gain	The Slider Gain is a Gain block for which the value of gain may be set using a slider control. Open the slider control by double clicking on the block. The slider can be moved during a simulation, thus providing a variable input device.

Nonlinear Block Library

Block	Purpose
lul Abs	Compute the absolute value of the input signal, which may be scalar or vector.
sin Elementary Math	The Elementary Math block can be configured to implement a number of mathematical functions, such as sine and cosine.

Nonlinear Block Library (Continued)

Block	Purpose
MinMax	The MinMax block can be configured to compute the minimum or maximum value of its inputs. The number of input ports can be set in the block dialog box.
Product	The Product block can be configured with one or more inputs. If there is one input, the output is the product of all elements of the input vector. If there are multiple inputs, the output is the element-by-element product of the vectors at each input port.
Combinatorial Logic	Implement a truth table.
Logical Operator	The Logical Operator can be configured to implement a number of logical operations, such as AND and OR.
Relational Operator	The Relational Operator block can be configured to implement a number of relational operations, such as less than or equal, greater than, etc.
Sign	Implement the signum nonlinearity. The output is 1 if the input is positive, 0 if the input is 0, and –1 if the input is negative.
Rate Limiter	Limit the rate of change of the output signal. When the input is changing, the rate of change of the output will be the same as the rate of change of the input, as long as the rate of change of the input is less than a settable limit. If the rate of change of the input exceeds the limit, the rate of change of the output will be the same as the limit.
Saturation	Implement a saturation nonlinearity. The upper and lower limits of the output signal are configuration parameters. If the value of the input signal is between the limits, the value of the output will be the same as the value of the input.

Nonlinear Block Library (Continued)

Block	Purpose
Quantizer	The Quantizer models an analog to digital converter. Its output is a multiple of the quantization interval, which is a block configuration parameter.
Coulomb & Viscous Friction	Implement a simple model of Coulomb and viscous friction.
Backlash	Implement a backlash nonlinearity.
Dead Zone	A dead zone is a region in which the output is zero. The upper and lower limits of the dead zone are configuration parameters. If the input is below the dead zone, the output is the input minus the lower limit, and if the input is above the dead zone, the output is the input minus the upper limit.
Look-Up Table	Perform linear interpolation in a table specified as a configuration parameter. This block maps a single input to a single output and maps a vector input to a vector output of the same dimension.
Look-Up Table (2-D)	The 2D Lookup Table maps two inputs to a single output.
Memory	The output of the Memory block is the value of its input at the beginning of the previous integration step.
Transport Delay	The Transport Delay block simulates a time delay. The output is the input delayed by a specified time.
Variable Transport Delay	The Variable Transport Delay block has two inputs. The first input is the signal to be delayed. The second input is the length of the delay. Thus the length of the delay can change during a simulation.

Nonlinear Block Library (Continued)

Block	Purpose
Hit Crossing	The Hit Crossing block causes the simulation to locate the instant (within machine precision) the input signal reaches a value that may be specified in the block dialog box. The block can be configured with or without an output signal. If there is an output signal, it will have a value of 0 except when the input is equal (within machine precision) to the value specified in the block dialog box. When the input signal is equal to the specified value, the block output is 1.
f(u) Fcn	The Fcn block may be configured with a function in a C language syntax. This block can accept a vector input but produces a scalar output. This block can't perform matrix arithmetic; however, it is faster than the MATLAB Fcn block, and is therefore preferable when matrix arithmetic is not needed.
MATLAB Function MATLAB Fcn	The MATLAB Fcn block implements a function using MATLAB syntax. It can accept vector inputs and produce vector outputs.
system S-Function	The S-Function block is used to incorporate a block written in MATLAB or C code into a SIMULINK model.
Switch	Switch between two input signals based on the value of a control signal.
Multiport Switch	The Multiport Switch can be configured to accept any number of inputs. A control signal determines which input is passed to the output.

Nonlinear Block Library (Continued)

Block	Purpose
f(z) Solve f(z) = 0 z Algebraic Constraint	The Algebraic Constraint block allows a SIMULINK model to solve algebraic equations. The model must be configured such that the input to the Algebraic Constraint block is dependent on the value of the output. A model containing an Algebraic Constraint block will attempt to adjust the value of the block output such that the value of the block input is 0.
Relay	Simulate a relay. The output is one of two specified discrete values, depending on the value of the input.

Connections Block Library

Block	Purpose
1 In	The Inport block creates an input for a subsystem. An Inport block can also be used to receive an external input (for example, using the `sim` command) to a model.
1 Out	The Outport block creates an output port for a subsystem. An Outport block can also be used to produce model outputs, to be used, for example, by the linearization or trim commands.
Mux Mux	The Mux block combines a configurable number of scalar input signals to produce a vector output signal.
Demux Demux	The Demux block splits a vector input signal into a configurable number of scalar output signals.
[A] From	A From block works with a Goto block. The output of a From block is the same as the input to the corresponding Goto block. A From block can receive input from only one Goto block, but a Goto block can send a signal to any number of From blocks.

Connections Block Library (Continued)

Block	Purpose
{A} Goto Tag Visibility	The Goto Tag Visibility determines which subsystems can contain From blocks corresponding to a particular Goto block.
[A] Goto	A Goto block sends its input to all corresponding From blocks.
A Data Store Read	The output is the current value of the contents of the corresponding Data Store.
A Data Store Memory	A Data Store Memory is a named memory location written to by Data Store Write blocks and read from by Data Store Read blocks. A simulation can save data in a Data Store Memory, and then access that data later.
A Data Store Write	A Data Store Write block writes to a specified Data Store Memory block. More than one Data Store Write block can write to a particular Data Store Memory, but if two or more Data Store Write blocks attempt to write to the same Data Store Memory on the same simulation step, the results are unpredictable.
Enable	Convert a subsystem into an enabled subsystem.
Trigger	Convert a subsystem into a triggered subsystem. Converts an enabled subsystem into a triggered and enabled subsystem.
Ground	Connect Ground blocks to unused inputs to prevent SIMULINK from producing error messages. For example, if a State-Space block is used to model the unforced behavior of a system, connect a Ground block to its input port.

Connections Block Library (Continued)

Block	Purpose
Terminator	Connect unused block outputs to Terminator blocks to prevent SIMULINK from producing error messages. For example, if you use a Demux block to split a vector signal but only need to use one component of the vector signal, connect the unneeded output ports of the Demux block to Terminator blocks.
[1] IC	The IC block sets the initial condition of its output to a specified value. After the simulation begins, the block output is the same as its input. This block is useful in algebraic loops, as it can provide an initial guess to the algebraic loop solver for the first time step.
Subsystem	A subsystem can be built in a Subsystem block as an alternative to encapsulating the subsystem using **Edit:Create Subsystem**.
Selector	The selector block accepts a vector input and produces a vector output that consists of selected elements of the input vector in a selected order.
0 Width	The output of the Width block is the number of elements in the input vector. Thus if the input is a five-element vector, the output is 5.

Index